FALLOUT

FALLOUT

DISASTERS, LIES, AND THE LEGACY OF THE NUCLEAR AGE

FRED PEARCE

Beacon Press, Boston

Beacon Press
Boston, Massachusetts
www.beacon.org

Beacon Press books
are published under the auspices of
the Unitarian Universalist Association of Congregations.

21 20 19 18   8 7 6 5 4 3 2 1

This book is printed on acid-free paper that meets the uncoated paper
ANSI/NISO specifications for permanence as revised in 1992.

Text design by Michael Starkman at
Wilsted & Taylor Publishing Services

Library of Congress Cataloging-in-Publication Data

Names: Pearce, Fred, author.
Title: Fallout : disasters, lies, and the legacy of the nuclear age / Fred
    Pearce.
Description: Boston : Beacon Press, 2018. | Includes bibliographical
    references and index.
Identifiers: LCCN 2017039844 (print) | LCCN 2017059975 (ebook) | ISBN
    9780807092507 (ebook) | ISBN 9780807092491 (hardback)
Subjects: LCSH: Radioactive pollution. | Nuclear accidents. | Nuclear
    industry. | BISAC: SCIENCE / Energy. | SCIENCE / Radiation.
Classification: LCC TD196.R3 (ebook) | LCC TD196.R3 P43 2018 (print) | DDC
    363.17/99—dc23
LC record available at https://lccn.loc.gov/2017039844

Contents

A Note on Units

In the nuclear world, units are a nightmare. There often seems to be a conspiracy to make them as complicated as possible, what with becquerels (and kilo-, mega-, giga-, tera-, and even petabecquerels), rems and rads and sieverts and curies and roentgens and grays and coulombs. What can all these different units be measuring?

Actually, most of the time, they are measuring one of two things. The first is the amount of radioactivity released in an accident, or perhaps present in a given amount of soil or water or air. The second is the radiation dose absorbed by a living organism, such as you. This second can be a bit more complicated because different kinds of radiation (alpha, beta, and gamma) from radioactive materials reach us in different ways. We might receive an external dose just by being in a radioactive environment or an internal dose by breathing in or eating some radioactive material.

Purists may be twitchy about my ignoring measures for "absorbed doses" and "exposure" and the like. But I have decided to keep things as simple as possible. I have chosen one unit for radioactivity and one for radiation dose, and ditched the rest of the statistical mumbo jumbo.

For radioactivity I have chosen the curie. This is an old measure, though still widely used. Some scientists prefer the becquerel. But this is such a tiny unit that we swiftly get into giga-, tera-, and peta- land. I hate that. It's like measuring a car journey in inches. There are a staggering thirty-seven billion becquerels in a curie. For most purposes, the curie is much more manageable. So curies it is. Just occasionally, when dealing with concentrations in small amounts, I have used the picocurie, which is a trillionth of a curie.

For radiation doses, I chose the modern measure of what radiologists call the "effective dose," which is weighted to reflect the different damage caused by different kinds of radiation. This is the sievert. It is a bit on the big side. A dose of just four sieverts will probably kill you. But the millisievert, a thousandth of a sievert, does good service. So I have used this through-out. Geiger counters will often measure doses in microsieverts per hour. Generally, and where it makes equal sense, I have converted them to mil-lisieverts per year.

As a rule of thumb then, a dose of four thousand millisieverts is usu-ally lethal; and one thousand millisieverts will probably give you radiation burns and perhaps a range of other potentially lethal symptoms known as acute radiation sickness. Below that, one hundred millisieverts is the low-est dose where there is reasonable evidence of a human health effect, such as more cancers in a population. For comparison, around two to three millisieverts is a typical annual dose from natural radiation sources; a mam-mogram is about the same. One millisievert is reckoned the maximum acceptable annual radiation dose for members of the public from power plants or nonmedical sources.

In other measurements, I have generally used US customary units: feet, pounds, gallons, and so on. I have converted euros, pounds sterling, and other currencies to US dollars. The exchange rate at the time of con-version made a euro worth $1.15 and a pound $1.30. The original units often appear in the end-noted sources.

Introduction

Anthropocene Journey

One sunny morning in September 1957, a line of military trucks drove down a narrow lane beside a lake in the foothills of the Ural Mountains, the chain that divides European Russia from Siberia. They stopped at a tiny village called Satlykovo. Red Army troops began knocking on doors and ordering the few hundred inhabitants to strip off their clothes, put on replacements unloaded from the trucks, and climb aboard. The villagers were being evacuated. They could not take any of their possessions, not even banknotes. As the evacuees bid a hasty goodbye to their worldly goods, the soldiers knocked down their homes to prevent them returning, and shot their cattle and pets.

The troops gave no explanation for the evacuation. They could not say—even if they knew—that a week before there had been an explosion in radioactive waste tanks at Russia's biggest plutonium factory, in the nearby closed city known today as Ozersk. Nor could they say that the strange dark cloud that had descended on Satlykovo hours later contained the deadly fallout from the accident.[1] Most likely it had been responsible for the death of Iran Khaerzamanov's ten-month-old daughter, who had been in the garden with her grandmother when the cloud descended. She suffered severe diarrhea and died hours later. Her body was the last to be buried at the village cemetery.[2]

The troops could not say any of this because the very existence of the nuclear-weapons complex was a military secret, known only to its fenced-in workers. Nobody outside was supposed to know—ever.

Sixty years later, on another bright, sunny morning, I became the first Western journalist to visit Satlykovo since the accident. I drove through a gate still guarded by armed troops and down a long rutted lane. I found

1

the village, but the remains of the seventy-five hastily demolished houses were consumed by vegetation. Nettles were everywhere. The hot, sticky air was thick with giant dragonflies. Across overgrown fields, the lake had plenty of fish, though nobody was allowed to catch them. The encroaching forest along the track harbored elk, wild boar, and foxes. Radioactive it may have been, but a barren wasteland it was not.

So started my journey to discovery the radioactive legacies of the nuclear age, an age I believe and hope is coming to a close. In this book, I explore the weird radioactive badlands created by nuclear accidents—some famous, such as Chernobyl and Fukushima, and some largely unknown, like the area around Satlykovo. I visit places where atomic bombs have been dropped, in the name of science or as acts of war, and where radioactive wolves roam but people fear to tread. I try also to make sense of our many personal and collective responses to the unleashing of the power of the atom, to the sense of foreboding and the all-too-real threat that it could be used to annihilate us all. In many ways this new psychological landscape turns out to be the strangest place of all.

||||

This is a personal journey too. I first became aware of nuclear technology during the Cuban missile crisis of 1962. I was ten years old and preparing to walk to school when my father told me out of the blue that if I saw a mushroom cloud during morning break that day, I should go back into the classroom and hide under my desk. I recall being confused, not least because I didn't quite know what a mushroom cloud looked like. Even so, during break I looked up at the sky for some time, just in case. I still remember the blue sky behind a big tree across the road, and a slight feeling of disappointment when no mushroom cloud appeared.

The realities of the nuclear age were new to everyone then: scary and somewhat secret, even to adults. At that age, I believed that adults knew everything; but about atomic stuff they were as much in awe and ignorance as we children. Growing up in southeast England, I would have been under an invisible radioactive cloud myself. It crossed the country after the fire at the Windscale plutonium plant in 1957. But nobody knew, because the cloud's path was a state secret. The nation was told the cloud went out to sea and wasn't very radioactive anyway. That was a lie, a lie told to adults as if they were children.

Like everyone else, my family back then was also paying for the privilege of living in the nuclear age. For the first couple of years of my life, every Brit had a ration card. It limited how much food we could buy. Rationing and austerity were what allowed bankrupt Britain to keep up with America and build its own atomic bomb. That felt good, except that having the bomb also made us sitting ducks if the Soviet Union—which got its atomic blueprints from the German-born British scientist and spy Klaus Fuchs—decided on a first strike. Or indeed a second strike.

Thankfully, nobody dropped a bomb in anger anywhere after the end of the Second World War. Only the Japanese suffered that fate as the war closed, at Hiroshima and Nagasaki. Even so, like the rest of the world in the 1950s and early 1960s, I breathed air infused with radioactivity from weapons tests in faraway places like Semipalatinsk, in Central Asia, and Bikini Atoll, in the Pacific. I also got to call the delightfully minimalist two-piece swimsuits worn by Brigitte Bardot and other exotics "bikinis." Because the truth was that, when I was young, the world was at least as much seduced as it was terrified by the atomic age. It was à la mode. And if it didn't blow us all up, atomic technology was going to transform our lives and, we were told, produce electricity that was "too cheap to meter."

That was then, but from the start this brave new atomic world carried the seeds of its own destruction. It hasn't been, I think, so much the radiation or the explosive power but the secrecy and deceit that have dogged it ever since. "Too cheap to meter" was always a lie, of course. So too was the myth that somehow "atoms for peace" had nothing to do with atoms for war.

I have in front of me a booklet produced in 1954 by the British government's Ministry of Supply—an Orwellian euphemism in itself, because it was actually a ministry of warfare—titled *Britain's Atomic Factories: The Story of Atomic Energy Production in Britain*. It has a picture on the cover of the two Windscale "piles" then manufacturing plutonium as fast as they could for British bombs. Yet the one-hundred-page booklet pretends for ninety-nine pages that the piles were designed to produce "limitless energy." They produced none. Only on the final page does it admit that their true purpose was to manufacture plutonium that "can be used as . . . the explosive in atomic bombs."[3]

Such duplicity, among governments across the globe, was routine. None of the tens of thousands of people drinking water from the tiny

Techa River in Siberia in the 1950s were told that the river carried lethal quantities of radioactive waste from a bomb-making plant a few miles upstream, or that their doctors were secretly cataloguing the health impacts. People near Denver, Colorado, believed the official line that the factory with the bright lights up on the hill called Rocky Flats was making household chemicals rather than plutonium "pits" for bombs. And certainly nobody told them that they had been showered with the radioactive metal after a fire there.

From the Windscale fire in the UK to the Chernobyl disaster, from Rocky Flats to Three Mile Island in the US, and from Mayak in the USSR to Fukushima, that compulsive secrecy, deviousness, and lack of accountability have persisted, even as the technology has morphed from military to civilian uses. Nuclear subterfuge has eroded public trust, debate, and decision making alike.

That explains why it has spawned some of the most trenchant environmentalism. Greenpeace had its origins in the Don't Make a Wave Committee, a group of British Columbians opposing US nuclear tests under the seabed off the Aleutian Islands in the Pacific in 1969. The German Green Party, the forerunner of many others round the world, was as much about opposing atomic technology on German soil as cleaning up the Rhine.

The bunker mentality of nuclear engineers has been matched by the hysteria of some antinuclear campaigning. The Don't Make a Wave Committee campaigned on the premise that the Aleutian tests would trigger an earthquake and a tsunami—something even the organizers admit they never thought would happen.[4] Anyone looking for the origins of modern "post-truth" politics could do worse than analyze its nuclear forebears. It is no wonder that most of us never believe what we are told about anything nuclear and always think the worst.

||||

However you view nuclear technology, it seems clear the world changed forever with the detonation of the first atomic bomb. From that moment on, the fate of the planet and of our species was in our hands—or more particularly in the hands of those people with their fingers on the nuclear button. It changed the way we think about many things. Globalization became a fact of life and death. No place was safe from "the bomb" and its fallout. The realization epitomized our new relationship with the Earth.

Scientists declare that we now live in the Anthropocene, a new geological era in which humans dominate the planet and its major processes, determining its climate, biodiversity, and chemical makeup. The body that decides these things, the International Commission on Stratigraphy, is expected to announce that the Holocene, which began with the conclusion of the last ice age, is now over and the Anthropocene has begun. The "primary signal" of this new age of humankind, it says, is the arrival of "plutonium fallout" from the first bombs.[5]

Plutonium is a new element, created by atomic scientists for its bomb-making potential.[6] Having been sprayed round the world by the fallout of bomb tests, it is now everywhere: bound to soils, in vegetation, and accumulating on the floors of the oceans. It is our unique signature on the planet and it will be around for millions of years, emitting its radiation as it decays. That dates the start of the Anthropocene to sunrise on July 16, 1945, when the first bomb containing plutonium was exploded over the desert of New Mexico. It was the dawn of a new age. It is no wonder nukes stir such emotions in us.

The atomic age plays to our guts and messes with our heads. Everybody has a view: for or against; optimist or pessimist; fearmonger or Panglossian technophile. It is not just because of the supreme power atomic energy brings. That is seductive or frightening, according to your temperament. There is also the radiation—unseeable, unsmellable, untouchable. It is like a ghost. We can make of it what we want, and we do.

Much of the time, we talk such nonsense about nuclear power. The power of science to control the atom seems to have unleashed irrationality in us. Both sides do it. Antis will hear no good; pros will hear no evil. Some antis attack the mainstream science of radiological risk with the same venom and disregard for truth as skeptics attacking the science of climate change. Pros respond with a technological hubris we rarely see elsewhere these days.

After researching this book, I have begun to wonder if we can ever get this right, or if there is something irretrievably dysfunctional about our relationship with nuclear technology. After more than seven decades living with the power of the atom, surely we should have learned to make sense of it by now. But all the evidence suggests not. The fallout in our heads shows no sign of going away.

Part One

The Destroyer of Worlds

THE ATOMIC AGE came very suddenly. Even some of the basic physics that underpinned it were unknown at the start of the 1930s. Yet by the end of 1945, two Japanese cities had been erased and bombs many times bigger were in preparation. Testing these "superbombs" through the 1950s rendered uninhabitable parts of Central Asia, the Pacific, the Arctic, Australia, and the American West—and left many injured and dead. Yet we got a thrill from this newness. The mushroom cloud became a symbol of hope as well as destruction. Nothing was more chic than to get up at dawn and watch the show in the Nevada desert. My journey through these first nuclear landscapes begins, as it must, in Hiroshima.

Chapter 1

Hiroshima

An Invisible Scar

I think I was expecting a big memorial at ground zero in central Hiroshima, at the spot beneath where the world's first atomic bomb exploded, blasting every building, triggering a firestorm, and vaporizing thousands of human beings. But the place of maximum destruction was marked only by a small plaque, set on a piece of marble the size of a parking meter and squeezed onto the sidewalk of a narrow street in front of a blank wall next to a carwash. I was there on commemoration day, seventy-one years after the blast. Someone had left a few flowers. As I watched from across the street, a family of Americans stopped for a couple of minutes to read the plaque. They took a selfie to send home. But everyone else walking down the street passed by without even noticing it.

Hiroshima seems studiously unmoved by the past. It has more than a million inhabitants, four times as many as on the day the bomb struck. Mazda's giant car factory in the suburbs has become a symbol of economic renewal; the Shinkansen bullet trains rush past on their raised tracks; the shopping malls are full of branded American goods. The streetcars still follow the routes they took in 1945, a ghostly memory of the past. But only five buildings in the entire center of Hiroshima survived the blast. Each has been preserved. One is the skeletal steel dome of the old Industrial Promotion Hall, beside the river. It is now a UNESCO World Heritage Site, and the most recognized symbol of what happened there. It is where tourists gather. A girl asked me to take her photograph, giving a thumbs-up and holding a fluffy toy.

I dropped into another, once the local headquarters of the Bank of Japan. The bulky classic stone building is just four hundred yards from ground zero. Today there are art exhibits inside. A small sign commemorates the building's survival. It notes matter-of-factly that "two days after the bombing, the bank was up and running in a limited capacity, and other financial institutions in the city set up temporary offices in the building." My jaw dropped at reading that. It seems that, even as the fires from the bomb continued to burn, as thousands of people were dying on the streets from radiation sickness, and bodies lay unburied, the business of money had resumed in Hiroshima. I love Japan, but I find its people very strange.

The world had seen nothing like it. On August 6, 1945, at 8:15 in the morning, an American B-29 bomber flew along the south coast of the Japanese island of Honshu. It dropped a single bomb, visible to the handful of people below who looked up at the plane and lived to tell the story, as a tiny black speck in the clear blue sky. The bomb—about ten feet long and code-named Little Boy—was about to change the world.[1] It fell for forty-five seconds toward the city of Hiroshima below. Then, at an altitude of two thousand feet—a height calculated by British mathematician William Penney to deliver the maximum destructive force—a gunpowder trigger fired a cylinder containing eighty-four pounds of uranium-235 into another containing fifty-five pounds of the same material. This collision created a nuclear chain reaction in which rapidly splitting uranium atoms released massive amounts of energy—equivalent to detonating thirteen thousand tons of TNT.

Right behind the B-29 came two more aircraft loaded with cameras and other equipment to monitor the devastation. It must have been a staggering sight. First came the white flash, bright enough to blind people on the ground. Then, within a fraction of a second, the heat from the explosion unleashed a fireball four hundred yards across that flashed to the ground below. There, temperatures of seven thousand degrees Celsius melted roof tiles and vaporized human flesh. Within seconds, all that remained of morning commuters was the shadows left on walls where their bodies took the initial heat. Behind the fireball came shock waves from the explosion that destroyed buildings and tipped over railcars, and a burst of radiation that killed many within hours.

The fires spread. Within twenty minutes, there was a firestorm two miles across, fueled by gas escaping from broken pipes. Dust and smoke shrouded the city. The inferno unleashed whirlwinds that ripped up trees in the city park, where many of those able to flee from the fires had sought shelter. Black rain fell: giant drops the size of marbles made up of radioactive soot suspended in water drops.[2] The fires burned for three days. The first burst of radiation from the bomb quickly disappeared. But the dust and debris on the ground near the blast remained radioactive for several weeks. It burned people who touched it, including rescuers, as well as killing fish in the Ota River and contaminating wells that people turned to for drinking.

Hiroshima that day had been home to roughly a quarter-million people, most of them living or working in the compact center over which the bomb had deliberately been burst. Some sixty thousand of them died on the first day, including 90 percent of those within five hundred yards of ground zero. Over the succeeding weeks, another forty thousand perished. Most were killed by the blast and fires that swept through the city. Perhaps they were the lucky ones. Others received horrendous doses of radiation—more than ten thousand millisieverts, assiduous American researchers later estimated—and died within a few days from internal bleeding or damaged organs and immune systems, a condition doctors termed acute radiation poisoning.

People with lesser doses died protracted deaths over weeks and months, often because their irradiated bodies were unable to manufacture new blood. People exposed to less than four thousand millisieverts usually survived, however, and their subsequent fate has been surprisingly good. Japan's Radiation Effects Research Foundation says today that anyone who received more than 150 millisieverts that day had an increased risk of leukemia or other cancers.[3] That was the majority of the city's inhabitants. But by 2000, there were only 573 more deaths than would have been likely anyway, or an extra 1 percent.[4] Also contrary to early expectations among scientists—and a continuing fear among the wider public—researchers have so far uncovered no signs of genetic mutations in succeeding generations. If there can be good news from Hiroshima, that is it.

||||

The bomb that dropped on Hiroshima came out of a clear blue sky both for the city and, metaphorically speaking, the world. There was no warning. There had been a few newspaper articles in which famous physicists such as Albert Einstein called for the development of nuclear weapons. Even so, before Hiroshima, few people knew that such weapons were even being developed, let alone being readied for dropping on a city.

It is hard to image today how, even in wartime, such secrecy could have persisted. For five years, the US government had been preparing the weapons as part of a huge clandestine industry. There had been one unpublicized test blast in New Mexico three weeks before. But on the day Hiroshima was hit, even the best-informed reporters knew little. News reports the next day were playing serious catch-up. In England, the *Manchester Guardian* began its story, "Japan has been hit with an atomic bomb 2,000 times more powerful than the ten-tonners dropped by the RAF in Germany . . . British and American scientists have been working on it for years."[5]

In the annals of war, the combined death toll of some 170,000 in Hiroshima and Nagasaki, the second city targeted, was not so exceptional. More than half a million soldiers had died at the Belgian village of Passchendaele during five months of bombardment at the height of the First World War. Probably fewer Japanese died in Hiroshima than had been engulfed in firestorms in Tokyo earlier in 1945 as American B-29s bombed the city relentlessly during two nights. Even so, the fact that one bomb aboard one plane out of a clear blue sky could destroy a whole city was terrifying.

The new reality of war has been made all the more compelling in the years since by the stream of stories from the survivors, known as the *hibakusha*. Most outsiders first read about them in the classic book published in 1946 by American journalist John Hersey, titled simply *Hiroshima*.[6] He followed the fate of six *hibakusha*. Since then, many others have told their stories in public—lest, they say, the world forgets. Several held a meeting during my visit to the city for the 2016 commemoration. No statistics can compare with the human stories I heard.

Take that told by Keiko Ogura, a short, bustling woman of seventy-nine. She was just eight years old and on her way to school when she was rendered unconscious by the blast. When she revived, she walked the city's streets dazed. She remembered passing hundreds of dead and dying, many with their burned skin peeling off. "What made me most scared was people grabbing at my ankles and saying, 'Give me water,'" she said. "I got some water from a well near a Shinto shrine and gave it to them. But as they drank, they vomited and died. Later, my father told me that it was bad to give water to people with burns. For years afterwards I had nightmares, and blamed myself for giving people water. I thought I had killed them. I didn't tell anyone till my father died. It was an invisible scar."

Kazuhiko Futagawa told how he was exposed to radiation in his mother's womb as she searched the blasted city for days. She was looking for her husband, who worked at the central post office just two hundred yards from ground zero, and for her thirteen-year-old daughter, who was creating firebreaks six hundred yards from ground zero. Thousands of suburban parents, he said, had lost their teenage children because the Hiroshima government had brought them into the city center that day to demolish buildings as a way of reducing the risk of firestorms if the city was bombed.

After the meeting I went to Hiroshima's modern memorial museum, built in a plaza in the city's peace park, which confirmed his story. Hiroshima, a port city on the estuary of the Ota River, was one of only two Japanese cities that had not been bombed by the Americans during 1945. (The other was the ancient capital of Kyoto, whose fabled temples one American general had visited and did not wish to see destroyed.) There were rumors—correct, as it turned out—that the Americans might be planning something "special" for Hiroshima. Some 8,400 of the city's teenage schoolchildren were working on firebreaks that day; 6,300 of them died.

The museum displayed several school uniforms left at home on the day the students went to create the firebreaks. There was the dress of first-year high school student Noriko Sado, donated by her mother, Mieko; and the trousers, boots, and hat of Koso Akita, a fifteen-year-old whose parents managed to get to him in the rubble before he died. I also found a school blouse donated by Futagawa, whom I had heard speak earlier. It had belonged to his older sister, who had been vaporized as she worked. He had wept as he told how he had found the blouse neatly folded in the back of a chest of drawers, after his eighty-seven-year-old mother had died. Throughout her life, he said, his mother had set her face against discussing the disaster that had befallen her family. The blouse, he said, "was the only thing left from her daughter, but she had never told us about it. I cannot imagine the suffering that lay behind that secret."

The museum's curators have shown a poignant regard for making sure every victim whose belongings are displayed is properly identified. So we know that the mangled tricycle in one room belonged to three-year-old Shinichi Tetsutani, whose father buried it with her but later had second thoughts and exhumed it for the museum. We learn that after the bomb dropped, Tsuneyo Okahara went to the office where her husband, forty-eight-year-old Masataro, had worked. She dug into the ruins in a desperate search for his remains. Eventually, on a desk, she discovered his lunchbox and ivory tobacco pipe—and bones in a chair at the desk. We are spared the bones, but the melted lunchbox and pipe were on display.

My day in Hiroshima had begun with the annual memorial ceremony, held in the peace park, created where the densely populated city-center neighborhood of Sarugaku-cho had once stood. The ceremony was a brief but somber affair, held at the hour the bomb had dropped. A peace bell was rung by bereaved families. There were short speeches by the mayor, the prime minister, and others; then the release of doves and, more democratically, the display of thousands of paper cranes made by participants, representing lost children.

Beyond the large tents housing invited guests, the crowds stood silently watching the events on TV screens, the sound of the music and speeches almost drowned out by cicadas in the trees. Many wore "No Nukes" stickers and fluttered fans with the same message. A few old people stood silently and remembered, their faces blank. Most were much younger, however. For them it was poignant history. Afterward, there were some, to my eyes, decidedly odd scenes of remembrance. A woman in a white evening dress sang in a quavering voice while her compatriot, also in evening dress, played a battered upright piano that had survived the blast. "The late Akiko's A-bombed piano" read the sign in English.

I was struck by how unflinching everyone was. Nobody tried to spare their children the horrors of the bomb. The memorial ceremony told of a boy who saw "charred corpses blocked the road. An eerie stench filled my nose. A sea of fire spread as far as I could see. Hiroshima was a living hell." Two sixth-grade students read a commitment to peace that began: "I smelled burning bodies. People's skin had melted. They didn't even look human." In the museum, parents showed their young children large images of burned and mutilated victims, and read out the signs carrying exhaustive lists of symptoms of radiation sickness: "Bleeding from gums and nose as mucous membranes failed, hair loss, bone marrow failure, low blood-cell counts, intestinal bleeding," and more. Some victims, the children learned, had literally coughed up their guts.

In my journey round the world exploring nuclear landscapes, I found that people often fantasized about the invisible effects of nuclear technologies. Here, where they had experienced the worst, there was a disturbingly brutal directness.

Equally striking—especially after watching the previous evening's TV reports of boorish antiforeigner sentiment during the 2016 US presidential election campaign—was how little anti-American rancor there was. None, in fact. Some expressed gratitude that President Barack Obama had visited the peace museum earlier in the year. Nobody, even when spoken to privately, had harsh words for the American tourists placidly viewing the commemorations. If the Japanese had dropped a similar bomb on mainland America, I could not image that Americans would be so uncensorious.

But there was something else. While the Hiroshima museum was very matter-of-fact about the bomb and its terrible effects, neither it nor the companion museum in Nagasaki mentioned that Japan surrendered a few days after the bombings, signaling the end of the Second World War. It seemed to me that Japan could contemplate the bombing with stoicism and dignity. The defeat, however, was too much.

As I left the museum, a woman staffing the exit approached everyone with a smile and a final message. "Please remember," she said, "that within a month of the bomb, the grass across Hiroshima started to grow again." Archive photographs behind her showed plants spreading rapidly along river banks and bursting through cracks in the roads, hiding the ash and invading the shattered buildings. It was "a message of hope," she said. I needed it as I passed the memorial mound, which contained the remains of seventy thousand people whose bodies had been discovered around the city, but who were never identified. Nature's resilience in the face of radiation became another motif of my travels through the world's nuclear landscapes.

Chapter 2

Critical Mass

MAUD in
the Nuclear Garden

Whatever its moral pitfalls, the production of the two atomic bombs dropped on Japan was a triumph of twentieth-century science. In the aftermath of Hiroshima and Nagasaki, the steam-powered industrial revolution suddenly seemed quaint. But the arrival of the new atomic age had been very sudden. It was the result of a tidal wave of new science about the structure of atoms, and how unstable these supposed building blocks of matter actually were.

This began with the discoveries early in the twentieth century that the apparently distinct atoms of each element—oxygen or uranium, copper or carbon—could take different forms, known as isotopes, and that these isotopes contained different numbers of neutrons, one of the building blocks of the atoms themselves. Most unsettling was the revelation that many isotopes were unstable. An isotope of one element might turn into an isotope of a different element, giving off radiation and energy as it did so.

What transformed this from fascinating science to a discovery that could change warfare was the insight that some atoms could be split to order, by bombarding them with neutrons released by another radioactive element. This splitting of the atom, or "fission," was first achieved in the lab by New Zealand physicist Ernest Rutherford, in 1917. But it was only in 1933 that Hungarian Leo Szilard suggested that it would be possible to set off an explosive chain reaction in which every split atom released many more neutrons that smashed into yet more atoms. At every step in this nuclear chain huge amounts of energy would be produced.

Szilard said the best chain reaction would come from splitting uranium, a metal with a large nucleus that would produce the most neutrons at each step. The uranium would need to be packed very tightly, so that a large proportion of the neutrons released by fission hit other uranium atoms. But if a "critical mass" of uranium could be brought together inside a bomb, it would explode with a force as great as setting off thousands of tons of TNT.

At the outbreak of the Second World War, most of the world's small band of nuclear scientists had fled from continental Europe and were working on these ideas in Britain and the US. At the end of 1939, Szilard, who was by now in America, met up with Albert Einstein, then the world's most famous physicist. They wrote a letter to President Franklin Roosevelt suggesting that even though America was then neutral in the war, it should develop such a bomb—not least in case the Germans were doing the same.[1]

Roosevelt initially didn't seem too interested. But in Britain, which faced the prospect of invasion by Germany, two other émigré physicists got a different response. A few weeks after Szilard's rebuff by Roosevelt, the Austrian physicist Otto Frisch and his German collaborator, Rudolf Peierls, wrote to Prime Minister Winston Churchill with the same idea. They added an important new detail. They had calculated that the critical mass of uranium needed to make a fission bomb was only twenty-two pounds, much less than most physicists had expected. There was a proviso: the bomb had to be made of one particular isotope of uranium, known as uranium-235, which made up only a small proportion of natural uranium.[2] But if it was, they promised Churchill that it would create an explosion that "would destroy life in a wide area . . . probably the centre of a big city."

This was the summer of 1940. The Battle of Britain was in full swing. The Germans were bombing London daily and an invasion might come at any moment. Within days, Churchill set up a secret committee, known for reasons never explained as the MAUD committee, to see how practical this proposal was—and how soon a bomb could be delivered.[3] Thus began the political process that ultimately delivered the Manhattan Project on the other side of the Atlantic, and the fateful dropping of two atomic weapons on already enfeebled Japan just five years later.

The MAUD committee quickly heard from another pair of émigré scientists, Austrian Hans von Halban and Russian Lew Kowarski. At their lab in Cambridge, they had been investigating how to generate useable electricity from the energy released by chain reactions. They figured that, rather than setting off a runaway fission explosion, it should be possible to control the chain reactions inside what they called a nuclear reactor.[4] This could generate energy useable for something other than destruction. But while investigating this, they realized that one of the products of splitting uranium atoms would be a new element, which they called plutonium. Plutonium did not exist in nature, but they calculated that one of its isotopes, plutonium-239, might be even more fissile than uranium-235. So even smaller amounts might make a bomb. In wartime, of course, nobody was interested in splitting atoms to generate electricity, but the idea of a plutonium bomb did grab the MAUD committee's attention.

Making the ingredients for an atomic bomb would require finding supplies of uranium ore and separating out uranium-235, or constructing reactors to make plutonium. Both were huge industrial projects and Britain didn't have the money. Only the Americans had the capacity to do the job.

Things briefly stalled until the US joined the war, following the attack on Pearl Harbor at the end of 1941. Then, Churchill ordered his scientists to share the conclusions of the MAUD committee with the American atomic elite. Within weeks President Roosevelt gave the green light to what became known as the Manhattan Project.[5] Soon, America was throwing hundreds of millions of dollars into turning the work of European university labs into the bombs to win the war.

The US government decided that in case one design didn't work out they would go full tilt to produce both a uranium and a plutonium bomb. By the end of 1942, a secret project was buying uranium from what was then virtually the world's only source, the Shinkolobwe mine in the remote far south of the Belgian Congo, and work was under way to extract uranium-235 from that ore.[6] Meanwhile, a nuclear chain reaction had been produced in a reactor in Chicago, and plutonium-239 had been extracted from it.

The Manhattan Project scientists developed a strange love-hate relationship with plutonium. Yes, it could destroy worlds, but it was also rather seductive. Thanks to its emission of radiation, it "feels warm, like a live rabbit," said Leona Marshall Libby, one of the few women scientists involved in the project.[7] Others reported that it had a metallic taste.

By mid-1943, a large expanse of remote sagebrush desert beside the Columbia River in Washington State had been commandeered for manufacturing plutonium-239, an isotope so fissile that a couple of pounds was thought capable of producing an explosion equivalent to twenty thousand tons of TNT. A huge construction enterprise employing thousands of workers at the Hanford reservation erected nine giant atomic reactors that bombarded uranium fuel with neutrons to create small amounts of plutonium. The fuel was then removed and the "spent fuel" was dissolved in nitric acid to extract the plutonium for turning into bombs, a chemical process called reprocessing.

The intellectual center of the Manhattan Project was far to the south at Los Alamos, a former boarding school in the New Mexico desert. Here hundreds of scientists spent their days drawing up blueprints for the bombs and for how to maximize their impact. Their average age was twenty-five. Almost all the British scientists who had contributed to the MAUD committee's report joined Robert Oppenheimer and other young US luminaries there. They included Peierls and Frisch, as well as their close colleague, the German-born mathematician Klaus Fuchs. Besides his day-to-day work, Fuchs kept up with everything. He had a photographic memory and, it later transpired, was sending all the secrets he learned to Joseph Stalin's chief nuclear scientist, Igor Kurchatov.[8] During his decade-long journey through the British and American atomic science establishments, this diffident but sociable and amenable émigré gathered up a massive amount of information.

Kurchatov soon knew that the Los Alamos scientists were designing both a uranium bomb and a plutonium bomb. In both bombs, the neutrons to start the reactions came from an "initiator" inside the bomb made of isotopes of polonium and beryllium. But otherwise the designs were very different. The uranium bomb used conventional explosives to slam together two small packs of uranium-235, creating the critical mass for a chain reaction. For the plutonium bomb, Oppenheimer and his colleagues decided they needed a more complex "implosion" bomb. There would be a single ball of plutonium about the size of a tennis ball. The critical mass would be created by detonating a shell of explosives around the ball to squeeze it. Calculating the physics of the implosion, and deciding exactly how to configure the explosives in the shell, were Fuchs's specialties.

The uranium bomb was never tested before being dropped on Hiroshima. But with more to go wrong, there was a test firing of a plutonium bomb in the New Mexico desert near Los Alamos in July 1945. It proved a dramatic success, with four times the anticipated explosive power. Just three weeks later, a duplicate bomb was dropped on Nagasaki. Days later, the Japanese emperor, Hirohito, surrendered. The job was done.

Spookily, the workforce involved in the Manhattan Project was officially put at 175,000 people, almost exactly matching the death toll from the two bombs.

||||

Many Manhattan Project scientists were fearful of what they had created. Robert Oppenheimer, their chief, invoked the words of the Hindu deity Krishna: "I am become death, the destroyer of worlds." There was anger, too, especially about the decision of the military to drop two bombs on Japanese cities. Szilard, the man who had originally conceived of the bomb, had argued for a demonstration of the new bomb's power in some remote location. But he was overruled by politicians and generals who wanted to see what would happen in a real city.[9]

With their deed accomplished, the scientists knew well that others could repeat it. To forestall a nuclear arms race, some called publicly for nuclear weapons to be put under international control. The generals didn't think much of that idea either. They rather liked the idea of being able to "destroy worlds." For a while after World War II, America hoped to keep the technology to itself. To that end, even their British scientific wartime collaborators were sent home—and rather ridiculously instructed not to use what they had learned should Britain decide to develop its own bomb.

Everyone overlooked Fuchs, however. He had for years made a point of keeping up with everyone's work. He shared what he knew with his British counterparts at Harwell in Oxfordshire, which quickly became Britain's equivalent of Los Alamos, and passed a constant stream of research papers and weapons designs to Kurchatov. Buoyed by the knowledge that the blueprints had obviously worked, the Soviet Union was able to accelerate its own weapons program. By the end of 1948, it was churning out plutonium at its own replica of Hanford—a hastily constructed closed nuclear city behind the Ural Mountains that is today known as Ozersk. The first Soviet plutonium bomb was tested on the steppes of Kazakhstan in August 1949.

An arms race had begun. But as Russia, the UK, and later France and others rushed to produce their own fission bombs, the Manhattan Project alumni had an even bigger bang in mind—what they at first called the "super-gadget." It was the brainchild of another Hungarian physicist, Edward Teller, and Polish mathematician Stanislaw Ulam. In its physics, the super-gadget was almost the opposite of the fission bomb. Fission bombs split heavy atoms such as uranium and plutonium. The new plan was for a "fusion bomb," which would force together light atoms of deuterium and tritium, both isotopes of hydrogen. Hence the popular name given to the new weapon: the hydrogen bomb. If Teller and Ulam were right, it would release many times the amount of energy of the fission bomb.

Triggering a chain of fusion reactions required a lot of energy, however. So much, Teller and Ulam concluded, that only a fission bomb could produce it. So each hydrogen bomb would have a fission bomb at its heart. Terrifying though the bombs dropped on Hiroshima and Nagasaki were, they were nothing compared with the new hydrogen bombs. The New Mexican desert could not contain its power and fallout. So the first true hydrogen bomb, Bravo, was exploded at Bikini Atoll in a remote region of the Pacific Ocean in March 1954. It had a force one thousand times larger than the Nagasaki bomb. It seemed there was no limit to how big a hydrogen bomb could be.

Early research into a fusion bomb was going on at Los Alamos even before the first fission bombs were dropped. Among the early developers was Fuchs. In May 1946, he filed a patent, which remains classified to this day, on how exactly to trigger a fusion bomb. It is not clear for whose benefit the Fuchs patent remains classified, since we now know that his Soviet minders had passed the details to Kurchatov and the "father" of the Soviet hydrogen bomb, Andrei Sakharov, almost as soon as it was written. When the archives of the Soviet Ministry of Atomic Energy were opened during the 1990s, the British military historian Mike Rossiter found copies of notes on the hydrogen bomb taken during lectures at Los Alamos in 1945, as well as reports of developments right through to 1948.[10]

During Fuchs's long journey through the British and American atomic science establishments, this genius had gathered up a massive amount of information. He spied for the British as much as for the Soviets. That may explain why, despite early suspicions across the Atlantic, his British bosses gave him free run of their nuclear establishments right up to the time he confessed his Soviet connections in 1950. He was subsequently convicted of espionage by the British, but was released after only nine years.

By the time Fuchs was put on a plane to East Germany, the mushroom clouds from Soviet hydrogen bombs were a regular sight on the steppes of Kazakhstan. And mushroom clouds had established themselves in the global psyche as the epitome of the new atomic age.

Chapter 3

Las Vegas

*Every Mushroom Cloud
Has a Silver Lining*

America's iconic nuclear landscape is the Nevada National Security Site, a fenced-off and largely deserted tract of sand, cactus, and Joshua trees that is bigger than Rhode Island. Once, when America was testing its atomic bombs here, it was the site of high jinks and revelry. Everything new and exciting in America was labeled "atomic," and Nevada was the place to experience the cutting edge of the new age.

The flashes could be seen 350 miles away in San Francisco. But in the up-and-coming desert resort of Las Vegas, less than seventy miles from the test site, the bombs were a weekend tourist attraction. The Chamber of Commerce tagged Las Vegas "Atomic City, USA" and distributed calendars giving detonation times. Staying up all night drinking atomic cocktails and then driving down Highway 95 for a closer look at the dawn blasts was the height of fashion. Or you could see the mushroom clouds and feel the ground shake from your hotel room. They charged premium prices for suites facing the test site.[1]

Even the stars felt the allure of the atomic. When a young Elvis Presley took the stage, Vegas billed him as "America's only atomic-powered singer." To add to the glitz, the city for several years crowned a Miss Atomic Bomb. Nuclear bombs, Elvis, and showgirls—what could be more Vegas? What could have been more emblematic of modern America?

There were four Miss Atomic Bombs. They reigned through the heyday of the desert tests, from 1952 to 1957. First was Candyce King, a dancer at Vegas's Last Frontier Hotel "radiating loveliness instead of deadly atomic particles," as one caption writer put it. Technically, she was Miss Atomic Blast, and there was no actual beauty contest. It was just a publicity shot of her wearing a mushroom cloud as a cap.

Next came Paula Harris, who sat on a parade float beside a mushroom cloud to depict the Oscar-nominated movie *The Atomic City*. Released in 1952, the film told the story of the kidnapping of a scientist's son in the secret bomb-making town of Los Alamos. She was followed in early 1955 by Linda Lawson, a singer at the Sands Hotel. She was said to have been crowned "Miss Cue" in ironic honor of the much-delayed Operation Cue, a series of blasts that year that tested the impact of atomic bombs on buildings, bridges, and other urban infrastructure.

Finally, and most famously, in 1957 there was another showgirl from the Sands Hotel who went by the name of Lee Merlin. She was photographed in a swimsuit largely consisting of a cotton mushroom cloud. That was the picture that did it. Blond curls in the breeze, arms spread high, red lips—and a white mushroom cloud. Oddly, to this day nobody knows what happened to her or whether that was her real name. She disappeared almost as quickly as the cloud itself.

So sexy was the bomb that, just as women got named after bombs, so bombs got named after women. A blast in June 1957—during which seven hundred pigs were deliberately exposed to massive radiation burns and flying glass to see how they got on—was called Priscilla. That was reputedly the name of a favored prostitute from Pahrump, a small town near the testing ground where many site workers were billeted.[2]

Kids were brought into the celebrations too. In 1954, St. George, a Mormon town in Utah downwind of the test that later suffered high cancer rates, crowned a young girl with a mushroom cloud on her skirt "Our Little A-Bomb." But bizarrely, says Robert Friedrichs, a radiation safety technician at the time who later researched the phenomenon for the test site's oral history project, the first Miss Atomic Bomb was not in Nevada at all. Not even in America. She was crowned after a beauty contest organized by the occupying US military forces in Nagasaki in 1946, just months after an American bomb had destroyed that city.[3] Pictures published in a women's journal of the day showed four finalists, all wearing kimonos rather than swimsuits, with a bunch of GIs standing behind them grinning.[4]

||||

Nevada was a latecomer to bomb testing. The first-ever detonation of an atomic bomb took place in secret at dawn on July 16, 1945. (Dawn was usually chosen for bomb tests because the air was still, so the spread of fallout would be minimized and the cloud would be as mushroom-shaped as possible.) The Trinity test took place from a 150-foot tower in the Jornada del Muerto desert, which translates as Journey of the Dead Man, south of Albuquerque in New Mexico. The detonation of a lump of plutonium the size of a tennis ball vaporized the tower and made a crater in the sand a thousand feet across. The cloud rose forty thousand feet above the desert. A roaring shock wave took forty seconds to reach the closest observers six miles away, many of whom were knocked to the ground. The sand around the crater melted to a green glass that geologists later named trinitite.[5]

In 1952, military engineers filled in the crater and erected an obelisk with a plaque. Today the site is part of the White Sands Missile Range, and is open to the public twice a year. Anyone tempted to pick up the flakes of trinitite glass still visible around the obelisk is told it is illegal to do so—and that the glass is still radioactive, flecked with plutonium that will take tens of thousands of years to decay.

After the war, bomb makers initially decided not to besmirch the American landscape with atomic tests. To conduct their continuing tests into ever larger bombs, they headed for the Marshall Islands in the Pacific Ocean, which had been recently liberated from Japan, and to one of its most remote atolls, Bikini Atoll. That's how we got the bikini. The first two-piece swimsuit began as the "Atome," excitedly marketed by French fashion designer Jacques Heim in early 1946 as "the world's smallest bathing suit." But after the first US atom test at Bikini that summer, a French automobile engineer named Louis Reard, who had just taken over his mother's lingerie business, brought out his own even smaller two-piece, named the Bikini. The Vatican called it "sinful"—not the bomb test, but the swimsuit.

After the Soviet Union went nuclear in 1949, the pace of testing heated up, and the convenience of the Nevada desert brought the atomic bombardiers back home. From January 27, 1951, when the ABLE "device" was detonated at Frenchman Flat, a dried-up lake bed in the middle of the new Nevada Test Site, the early-morning skies were regularly illuminated by the tests, which often received live national TV coverage.

That's when the whole nation became enthralled by the atomic spectacle. Everything from clocks to lamps to corporate logos soon adopted "atomic" designs, such as a mushrooms cloud or the nucleus of an atom circled by electrons. High school football teams were renamed the Atoms. (One school team near the Hanford plutonium complex still has a mushroom cloud as its symbol.)[6] The thrall was spiced with fear. This was the McCarthy era, when public hearings chaired by Senator Joe McCarthy into suspected Communist infiltration of the government led to a period of political paranoia. But there were real spies, too, such as the recently imprisoned Fuchs. And the fear of an all-out nuclear war between America and the Soviets led to scarily methodical preparations.

Spookier than the rubbernecking nuclear test observers in their Vegas penthouse hotel suites were the "survival towns." Around the test sites, soldiers constructed replicas of 1950s suburban America, complete with fully furnished homes occupied by mannequins. Their destruction was filmed in minute detail to see what would happened if the land of motherhood and apple pie were blown to smithereens.[7]

Not far away, also within the testing grounds, was another town that was almost as surreal. Mercury, an old mining community five miles off Highway 95, became America's very own closed nuclear city. It was the gateway to the testing ground but was cut off from the outside world by military guards. At one point, Mercury had a population of more than ten thousand employees, rivaling the permanent population of Las Vegas. It had its own bowling alley, movie theater, swimming pool, chapel, clinic, library, and Atomic Motel—not to mention the Desert Rock Air Strip, built especially so President John Kennedy could land there in 1963. Mercury was a full-scale replica of suburban America—the one they did not blow up. It is still there. Though much diminished, and with a population reduced to just five hundred souls, this time capsule from the 1950s lives on.

IIII

There were around a hundred atmospheric tests in Nevada from 1951 to 1962. Any fears that witnessing the tests might be dangerous were assuaged by brochures from the Atomic Energy Commission. One reassured locals and sightseers that even though "some of you have been exposed to potential risk from flash, blast or fallout," the radiation from the tests was "only slightly more than normal radiation . . . wherever you may live" and "does not constitute a serious hazard to any living thing outside the test site."[8]

In any case, being a witness was patriotic. "You are in a very real sense active participants in the nation's atomic test program," the brochure said. Well, yes, but the witnesses were also unwitting guinea pigs in a national experiment. As they spread on the winds, the mushroom clouds rained radioactive debris across the whole country. Statisticians would in later years pore over health records to see how dangerous this fallout might have been. Many citizens went to court demanding payouts to compensate for the harm they claimed they had suffered.

Local residents began asking questions early on about whether the tests were as safe as the Atomic Energy Commission insisted. In early 1953, around four thousand sheep, including newborn lambs, died as they grazed on pastures fifty miles downwind of the test site. It was hushed up at the time. Dead lambs were bad PR, but two decades later, researchers concluded, despite continuing denials from the commission, that the sheep had died from eating grass made radioactive by the fallout. A few weeks after the dead-sheep episode, official alarm was such that roads over the state line in California were shut, and dozens of vehicles decontaminated, after the fallout from another test called Simon grounded soon after release. In 1958, a fallout cloud got caught up in Los Angeles smog.[9]

Scientists played catch-up with public fears. First they looked at service personnel. A study in the late 1970s found that the three thousand troops who had been lined up to witness the Smokey test in August 1957 had subsequently suffered twice the expected number of leukemias. Was that a coincidence, or a result of more assiduous counting, or a real effect? All three explanations are possible. The American Cancer Society later claimed to find leukemia rates on average three times higher among service personnel who witnessed the tests in the line of duty; but other studies found no such link.[10] With widespread compensation payments made in advance of scientific proof, the push to resolve the truth has diminished. The links remain unproven.

Then researchers looked at civilians living downwind of the tests. In the 1980s, Carl Johnson, an epidemiologist from Colorado, found cancer rates 60 percent higher than expected among four thousand Mormons living in the small towns of southwestern Utah. They included St. George, which crowned "Our Little A-Bomb" girl in 1954 and is less than 150 miles downwind of the Nevada Test Site.[11] St. George loved the bomb and certainly put its citizens in harm's way. Schools regularly bused their students to vantage points to watch the mushroom clouds form to the south.

The town took a direct hit in May 1953 after a change in wind direction dumped fallout from a bomb called Harry—later dubbed "Dirty Harry" after the discovery that its fallout exceeded that of any other atmospheric test in the continental US.[12] Individual radiation doses that day were estimated at 60 millisieverts, more than three times the then-limit for members of the public. The radiation biologist doing the measurements later told a court that he had been instructed to deny the doses measured. He claimed his official report was tampered with.[13] The sixty-millisieverts dose is still below the level at which cancers and other harmful effects have been convincingly demonstrated, which is generally held to be around one hundred millisieverts. But the people of southwestern Utah had been exposed to many tests. A radiation expert from the Atomic Energy Commission, John Gofman, told a district court in 1984 that he estimated the cumulative dose of people living in the area from fallout through the 1950s was around 360 millisieverts, well within the range that would likely cause harmful effects.[14]

Partly on the basis of Gofman's evidence, the judge ruled that the fallout had caused ten people in the state to die of cancer, including two thirteen-year-olds: Sheldon Nisson, of St. George, and Sybil Johnson, of nearby Cedar City. The judge also found the government guilty of negligence in the way it had conducted the tests and had failed to warn "downwinders" about the dangers. The legal ruling was overturned by an appeals court and later by the Supreme Court, but in each case only on the grounds that the law did not cover national security activities.[15]

Many have suggested that victims of the Nevada tests might have included the actor John Wayne. He was on location near St. George in 1974 making *The Conqueror*, a critically panned film about Genghis Khan. That was long after the end of the tests, but Wayne would have been kicking up sand hit by fallout. He died of stomach cancer five years later. Three of Wayne's costars—Dick Powell, Susan Hayward, and Agnes Moorehead— also died of cancer, while ninety-one of the 220-strong cast and crew later contracted cancer and forty-six died of it.[16] But with half of Americans then developing cancer in their lifetimes and a quarter dying from the disease, the case is, at best, unproven.

After the US and other nuclear nations signed the Partial Test Ban

Treaty of 1963, atmospheric tests in Nevada ceased. But underground tests continued in boreholes and tunnels. Such tests briefly vaporized and melted rocks but generally didn't release radiation to the atmosphere. There were exceptions, however. One test in shallow ground, the Sedan test, created the country's largest man-made crater, 1,200 feet wide and 300 feet deep, and released more radioactive contamination across the US than any of the nation's atmospheric tests. More even than Dirty Harry.[17] Nobody told the downwinders to take cover.

Unless you fear you are living with a radiation-induced disease, all this must feel like long ago. More than half a century has passed since the last mushroom cloud rose over the Nevada desert. Even so, if you drive out along Highway 95 from Las Vegas, you will see signs of more modern warfare. Just before you reach the turning for the "time capsule" town of Mercury, you pass the Creech Air Force Base. Here, Air Force "pilots" sit in temporary cabins all day, operating the US's growing fleet of combat drones, as they make killing expeditions in war zones thousands of miles away in Syria, Afghanistan, and Somalia.[18]

Las Vegas is still booming, of course, but the allure of the bomb that once helped fuel its growth has faded. Back in 2016, some bright London producer came up with a West End musical comedy titled *Miss Atomic Bomb*—set in 1952 in Vegas, "where every mushroom cloud has a silver lining and fallout is your friend." It must have seemed like a surefire winner: nostalgia with a twist. But it didn't do much business and closed within weeks. Even the Bikini Martinis for sale in the St. James Theatre bar fell flat.

Chapter 4

Pacific Tests

Godzilla *and* *the* Lucky Dragon

It was just before dawn on March 1, 1954. A Japanese trawler called *Daigo Fukuryu Maru*, which means *Lucky Dragon No. 5*, was fishing for tuna more than 1,200 miles from home amid the Marshall Islands of the tropical Pacific Ocean. As the lookout awaited the rise of the sun in the east, he was astonished to see instead a flaming orange ball suddenly appear in the other direction, on the western horizon. He called the rest of the twenty-three-member crew on deck and they watched in consternation as the "sun" faded and a huge mushroom cloud formed in the sky.

Two hours later, as they hauled in their nets, white ash began to fall on their boat. The fallout continued for five hours. The bemused fishers rubbed the mysterious sticky ash between their fingers and one of them gave it a lick. After a while, they noticed the ash was so thick that they were leaving footprints on the deck. Though they didn't know it, the white ash was the radioactive remains of the coral island of Namu on Bikini Atoll one hundred miles away. Its ring of coral had been blasted to pieces by the detonation of the Bravo hydrogen bomb, America's first true hydrogen bomb. With an explosive power of fifteen megatons, a thousand times bigger than the Hiroshima bomb, it is to this day the largest nuclear bomb America has ever detonated. The "sun" had been a fireball four miles across created by the explosion.[1]

As the fishers headed for home, they began vomiting over the side. Radiation burns and sores appeared on their skin. Their hair started to fall out. Their gums began to bleed. When they docked at Yaizu in southern Japan two weeks later, they were no better. While some went home, others headed for the city's hospital. With the medical conditions from the Hiroshima and Nagasaki bombs fresh in their minds, doctors quickly figured out what had happened. The sailors were radioactive and suffering from acute radiation poisoning. The rest of the crew were rushed to the wards.

The Japanese sailors received between two thousand and six thousand millisieverts of radiation, a potentially lethal dose.[2] However, when one of the doctors in Yaizu wrote to the US government's Atomic Energy Commission asking how to treat the victims, he did not get a reply. The bomb bureaucrats feared any information they gave about which isotopes were in the fallout would reveal secrets about the bomb's makeup. Instead, commission director Lewis Strauss—a self-made millionaire and ardent anticommunist famous for having got Manhattan Project boss Robert Oppenheimer's security clearance revoked—let it be known he thought the tuna fishermen were a "Red spy outfit."[3]

In fact, the crew of *Lucky Dragon No. 5* were simply in the wrong place at the wrong time. When Bravo exploded, they were well outside the official danger zone. But the wind had shifted just before the explosion, leaving them in the direct line of the fallout. It soon emerged that as many as a hundred Japanese tuna fishing vessels had been in harm's way. In the ensuing panic, tuna was removed from the shelves of fishmongers across Japan. It ignited a concern in Japan about nuclear bombs that had lain dormant since the trauma of Hiroshima and Nagasaki. The Japanese film *Godzilla*, made in response to the tragedy and released seven months later, proved hugely popular. It described that country's fictional efforts to protect its citizens from a marauding mutant monster created by atomic tests on a Japanese island.

While the nation binged on atomic horror, the crew were still in hospital receiving regular blood transfusions as doctors tried to decontaminate them. One, the radio operator Aikichi Kuboyama, died of a hepatitis infection arising from the transfusions. The others lived, several into their eighties, though they suffered years of being stigmatized by their friends and neighbors, many of whom believed radiation sickness was catching.

||||

Bravo marked the start of an era of testing bombs that were far bigger than those dropped on Hiroshima and Nagasaki, and often far too big to be detonated in the Nevada desert or even in the Soviet Union's favored testing grounds on the steppes of Kazakhstan. For these megaton explosions

America—and later Britain and France—chose apparently remote Pacific atolls, thousands of miles from major centers of population. The Soviet Union chose the Arctic. But in truth, nobody was safe anymore. Or that was how it seemed. A single hydrogen bomb could now destroy any city in the world—including New York, as Strauss boasted to one American interviewer. And its fallout was global, because the debris rose much higher into the air.

A nuclear explosion creates a huge fireball that propels into the air everything within it—including soil, water, and, in the case of Bikini, the blasted remains of the coral reef where it was detonated. This cloud of debris eventually loses its upward momentum and spreads out, creating the iconic shape of a mushroom cloud. The cloud then disperses on the winds and gradually falls to Earth. Some of this fallout reaches the ground within a few minutes, as the *Lucky Dragon* fishermen discovered. But the material that gets blasted high into the atmosphere can stay aloft for many months.

The bigger the explosion, the higher the cloud takes the debris, the longer it takes to come down, and the farther it spreads. All the debris from the early atomic bombs fell to Earth within seventy days. But Bravo's fireball rose to forty-five thousand feet, higher than a cruising airliner. Its radioactive cloud broke clear through into the stratosphere, reaching an astonishing 110,000 feet. Much of its contents took eighteen months to fall to Earth. This meant that some of the more virulently radioactive isotopes with shorter half-lives had decayed before reaching the ground but it also meant that the fallout spread globally, covering the planet in a thin radioactive smear. It delivered the highest fallout doses of radiation in the history of worldwide nuclear testing.[4]

Anger at this radioactive violation of humanity spread round the world. Fears grew that a war would deliver a nuclear holocaust that destroyed all life on Earth. Nevil Shute's novel *On the Beach*, published in 1957, told the story of a group of Australians awaiting death from the arrival of fallout created by a nuclear war that had killed everyone in the northern hemisphere.[5] Humanity for the first time became aware that it faced an existential threat of its own making. But for some communities in the Pacific that threat was immediate.

‖‖‖

The coral atolls of the Marshall Islands are among the most remote places on the planet. When the US took control of the archipelago from defeated Japanese soldiers at the end of the Second World War, generals swiftly identified two of the atolls, Bikini and Enewetak, as suitable for bomb tests. The first atomic weapons tests after Hiroshima and Nagasaki were conducted there in 1946, before the urgency of the arms race encouraged relocation to Nevada in 1951. Then, from 1954 to 1958, the islands hosted the far more violent hydrogen bombs, with a total explosive power of 109 megatons, some seventy-five times that visited on the Nevada desert.[6]

To make way for the first tests, Commodore Ben H. Wyatt, the military governor of the Marshall Islands, went to see the 167 inhabitants of the tiny necklace of islands around the Bikini lagoon in 1946. He "invited" them to leave, telling them it was "for the good of mankind and to end all wars."[7] They believed what they were told by their recent liberator, and expected they would be home soon. As their traditional leader Dretin Jokdru put it before his death in exile in 2006: *"They told us . . . 'Never mind if you are living on a sandbar, we will take care of you as if you are our very own children'. . . . In a way we were happy that they would be taking care of us. The world was a strange place for us then. We just couldn't understand why they wanted our island."*[8]

The manner in which they were "taken care of" came as a shock, however. They were dumped on Rongerik Atoll 150 miles to the east, where they nearly starved before being moved into tents beside the runway at a US base on Kwajalein Island, and then on again to Kili Island, a place without a lagoon, so they could not even fish in their traditional way.

At least they had been evacuated. When Bravo exploded, eighteen residents of Rongelap Island were gathering copra on an atoll, just ninety miles away. Like the Japanese fishers, they were doused in radioactive coral dust. Two hours later, the fallout hit Rongelap itself and showered another sixty-eight residents in their homes. The next day, American servicemen in protective suits showed up and took some measurements. By then, like the fishers, most residents were suffering burns and vomiting— clear sign of radiation sickness. They were hurriedly evacuated to Kwajalein, after which the Atomic Energy Commission put out a press statement declaring that the evacuation was "according to plans as a precautionary measure. . . . There were no burns. All are reported well."[9]

After atmospheric tests ceased in 1963, America had no further use for the islands and was keen to repatriate the evacuees. In 1968, President Lyndon Johnson declared Bikini Atoll safe, and four years later more than a hundred people returned from Kili. Subsequent health checks found their bodies contained raised levels of radioactive cesium-137 from the bomb fallout, however. So they were evacuated again.[10]

Cesium-137 has a half-life of thirty years, meaning that half of it will have decayed away in that time. So by now, radiation levels should be not much more than a quarter of what they were after the tests. But even in 2016, a study concluded that Bikini still "may not be safe for habitation." The remaining fallout in the soil concentrates in the coconuts and the crabs that live on the coconuts, it said.[11] Some say that the radiological hazards are exaggerated. One study puts the dose from a diet of local food on Bikini at fifteen millisieverts a year. That is six times the global average natural background but well below any level demonstrated to cause health problems. Arguably, the psychological complications of exile are worse than the radiological hazards of returning.[12] Even so, rather understandably, the people from Bikini simply don't trust Americans who say it is safe to return. They still feel they are being treated as children.

The chaos has been even worse for the fallout-showered people of Rongelap, evacuated after the Bravo blast. In 1957, just three years after Bravo, the Americans shipped them back to their island, without any effort at decontamination.[13] There they remained for three decades, mostly living off their poisoned land. But after experiencing a rising number of thyroid cancers among their children and leukemia cases among adults, they pleaded to be removed again. Greenpeace sailed to the rescue and shipped them to Kwajalein. There they suffered from overcrowding, a lack of jobs, and a spate of suicides.[14]

In 1998, the US government finally began a program of treating Rongelap's radioactive soils. Since then, some people have returned and the community reportedly thrives again. It has a paved runway, hosts eco-tourists, and sells locally grown black pearls.[15]

In addition to Bikini, the US conducted forty-three tests on the evacuated Enewetak Atoll. In 1979, its cleanup crews arrived. They scraped up some 3.5 million cubic feet of the most radioactive soil and dumped it all into a concrete-capped sarcophagus known as the Runit Dome. The promise is that, with further decay of isotopes like cesium-137, most of the atoll will be safe for humans by the late 2020s. But the Dome, which looks like a half-buried UFO sticking out of the sand, is already cracked and may be leaking its contents back into soils, according to a follow-up survey by the Department of Energy in 2012.[16]

These are sorry tales. No outsiders come out of it well. The islanders have too often become pawns in a wider battle for hearts and minds over the risks of radiation.[17] Their plight has been complicated by the failure to give proper compensation. While a billion dollars had been given to US citizens exposed to bomb tests, less than a tenth as much has gone to the Marshallese, many of whom received much higher doses.

But what of nature? Have the islands become a radiological death zone? The smashing of coral reefs and the contamination of soils and lagoons during the tests were substantial. But, like many other radioactive places round the world, nature is recovering and may even be prospering in the absence of humans. When, in 2008, marine biologists dived into the underwater crater created half a century before by Bravo, they found new coral reefs growing up to twenty-five feet high. The reef was being recolonized by coral from neighboring atolls. "I didn't know what to expect, some kind of moonscape perhaps. But it was incredible," Zoe Richards, of Australia's James Cook University, told Reuters.[18]

On the bottom of the Bikini lagoon lies an armada of old warships deliberately sunk by the US military prior to the Bravo blast to test the destructive power of its new weapon. Among them, with a symbolism that is easy to fathom, is the *Nagato*, the Japanese flagship from which Admiral Isoroku Yamamoto gave the order to attack Pearl Harbor. In 2010, UNESCO declared Bikini a World Heritage Site, because it "symbolises the dawn of the nuclear age."

||||

The British conducted their first atomic-bomb tests in Australia but then sought out Pacific islands for hydrogen-bomb tests. From among their handily remote Pacific imperial possessions, they chose the Christmas and Malden Islands, part of the now-independent South Pacific state of Kiribati. The British carried out many fewer tests than their American and Soviet rivals, and so wanted to collect every last scrap of data from each test. That was not good news for the service personnel or locals conscripted to help.

The biggest explosion, Grapple Y over Christmas Island, yielded three megatons. Airmen were nonetheless told to fly close to the explosion, right through the mushroom cloud, to collect samples of radioactive fall-out. Being told to shut their eyes during the blast didn't help much. One reported "seeing the light through our eyelids. It was just incredible. It blew our minds away."[19] Those not sent into the air were lined up on the beach and told, according to Kenneth McGinley, who was nineteen at the time, to "clench our fists and push them into our eye sockets" while the bomb was detonated twelve miles away.[20] Afterward, they ate local fruit, drank local water, and bathed in the lagoon. Before going off duty, they were ordered to kill wild birds that had been blinded by the explosion.[21]

Grapple Y caused radioactive rain over Christmas Island, where the residents remained throughout. One, Suitupe Kirotomi, remembered "looking up at the black cloud from the blast, which was directly above us when the light shower fell." Later, a red burn appeared on her face and her hair began to fall out. "The mark remains on my face today," she told reporters in 2006.[22]

How much harm this unexpurgated exposure did to the troops and islanders remains contested. In 2015, the Fijian government compensated seventy of its people who were there. But studying fifty New Zealand servicemen who took part in the Christmas Island tests, Al Rowland, of Massey University in New Zealand, found the cancer data inconclusive.[23] On the British side, McGinley said he suffered recurrent skin problems and sterility. He founded the British Nuclear Test Veterans Association to push for compensation.

The British government has continued to argue that servicemen have not shown sufficient evidence that they were harmed. It may have a scientific case, but it is hard not to be cynical when the government mandarins seem to have deliberately avoided collecting such evidence, despite being warned by government scientists at the time that this might land them in trouble. The minutes of a meeting held at Britain's Atomic Weapons Establishment in Aldermaston in July 1958, a month before one of the Grapple tests, reveal that Aldermaston scientists wanted to take blood samples from servicemen before they took part in the tests. This would allow them to check whether white blood-cell counts subsequently fell, which would be a possible prelude to leukemia.

The minutes record that the scientists were also "concerned about the political repercussions which might ensue if charges of negligence, however unfounded, could be proved. It would prejudice the case if no blood count was taken and a person became ill later."[24] Military people at the meeting turned down the proposal. The feared repercussions played out forty years later when the European Commission on Human Rights ruled that the British government had acted illegally and dishonestly toward its servicemen.[25]

The French initially carried out atmospheric tests in the deserts of Algeria. After that country won its independence, they too moved to their own Pacific islands. Between 1966 and 1974, their forty-one atmospheric tests at the uninhabited Mururoa and Fangataufa atolls showered fallout across much of Polynesia. This was in defiance of the 1963 international ban on atmospheric tests, which France had refused to sign. Like the Americans and British before them, the French regarded the locals primarily as an inconvenience, displaying persistent secrecy, denial, and bluster at the suggestion that the fallout might pose a threat.

In the first year of operation, radiation was measured on the Gambier Islands, 250 miles from Mururoa, at five times the permitted annual dose. None of the several hundred people there had been evacuated. The following year, two French meteorologists stationed seventy miles away were hospitalized after being showered with fallout. Yet nobody checked the sixty inhabitants of the atoll.[26]

After ending atmospheric tests, the French continued with underground tests. In 1985, after a Greenpeace ship, the *Green Warrior*, invaded the testing zone, government agents got tough. One night they blew up the boat in a New Zealand harbor, accidentally killing a photographer who had returned to the ship. Tests ended in 1996, leaving behind an atoll that is fractured and riddled with radioactive material from underground tests, including around forty pounds of plutonium, which fell into the lagoon after a bomb broke apart.[27]

||||

Since 1945, more than five hundred atomic weapons have been tested in the atmosphere at thirteen main sites, with a cumulative explosive force of 440 megatons, equivalent to a staggering twenty-nine thousand Hiroshima bombs. Their total release of radioactive fallout was about six hundred times that from the Chernobyl accident in 1986, the worst nonmilitary disaster. Of those 440 megatons, more than half, with a total force of 239 megatons, were inflicted on the mountainous Russian Arctic island of Novaya Zemlya.

It was there that the world's biggest-ever weapons test was exploded, on October 30, 1961. At fifty megatons, the Tsar bomb was more than three times bigger than the largest US bomb. It had ten times the explosive force of all the weapons used in the Second World War, and its mushroom cloud grew to seven times the height of Mount Everest and measured sixty miles across. The heat caused third-degree burns seventy miles away and window panes shattered six hundred miles away. Mankind has never done anything else like it. Unleashing a quarter as much energy as the Krakatoa volcanic eruption, it was more like a geological event than a bomb. Much the same could be said of the 133 underground tests carried out at Novaya Zemlya after 1963. One triggered a landslide that blocked two glaciers and created a lake more than a mile long. Another melted permafrost to a depth of more than three hundred feet.[28]

Everybody on the planet got some of the fallout from the decade of giant tests between 1954 and 1963; nobody was spared. The fallout is detectable in tree rings and coral reefs, soils and coastal sediments on every continent. The average human picked up a dose of 0.15 millisieverts in the peak year of 1963. That was only about 4 percent of a typical annual dose from natural background radiation. Only in a few areas were doses likely to have been sufficient to cause harm, though as we will see in chapter 20, there is huge uncertainty about how many people among billions might be killed by tiny doses. The fallout could have killed nobody, or it could have killed tens of thousands.

But while such global questions are obviously important, we should not lose sight of the many more immediate problems that have engulfed people who had the misfortune to be the neighbors of bomb tests. To understand more of that, I have combed the fragmentary but increasingly compelling evidence of what happened around the Soviet testing grounds at Semipalatinsk, on the steppes of eastern Kazakhstan, where the winds often brought with them a lethal radioactive mist.

Chapter 5

Semipalatinsk
Secrets of the Steppe

"For many years, this has been a secret," wrote Kazbek Apsalikov, director of the Institute of Radiation Medicine and Ecology (IRME), an agency of the government of Kazakhstan. He was responding to an e-mail I sent him after reading a report he had cowritten on the Soviet nuclear bomb tests in his country at Semipalatinsk half a century ago. The 2014 report had referred, in a single intriguing paragraph, to a deadly cloud of fallout that had engulfed a major Kazakh city in 1956. The quarter-million population of Ust-Kamenogorsk, a hub of heavy industry 250 miles downwind of the blast, "was exposed to nuclear fallout with radiation doses so high as to cause acute radiation poisoning," Apsalikov had written. As a result "638 people suffering from radiation poisoning were put in a special hospital—Dispensary No. 3 . . . Moscow sent a 'special commission' headed by a minister to find out what had happened."[1]

Apsalikov had himself stumbled on this intriguing snippet of information in the once-closed archives of his institute. "Unfortunately, no further information about the fate of these people is available," he told me. He believed most of the files had been taken to Moscow, either at the time or when the Soviet Union broke up in 1991. What happened that day was meant to remain a secret. We still don't know, for instance, how many of those hospitalized with acute radiation poisoning eventually died. We do know that the number of cases of poisoning was far higher than the 134 who suffered similar symptoms after the Chernobyl accident, of whom twenty-eight soon died. Pro rata, 130 would have died in Ust-Kamenogorsk. But that is just a guess.

After opening in 1949, Semipalatinsk had become the busiest bomb testing ground in the world. In all, there were 619 tests, 122 of them atmospheric. It was also the most secretive. Nobody downwind was told anything about what the bomb-testers were up to. Many felt the ground shake and saw the flashes and mushroom clouds, but it didn't do to ask too many questions in the days of Joseph Stalin. Especially in Ust-Kamenogorsk, a closed city that both mined and processed uranium for the nuclear industry.

Even so, the descent of a fallout cloud on such a large city proved a wake-up call for the Soviet authorities. This, they would no doubt have thought, was not a case of a few expendable peasants caught in harm's way.[2] Hence the "special commission," largely comprising scientists from the Institute of Biophysics in Moscow, called for research to measure radioactivity and establish the doses received by the million or more people living downwind of the testing grounds, known to the Soviets as the Polygon.

The hastily established hospital known as "Dispensary No. 3" turned into a permanent research facility and became IRME, which Apsalikov now runs. It has charted the victims of the tests ever since, initially as a secret Soviet research institute and since 1991 as a public body. Before 1991, little of what it learned was passed on even to local scientists. After 1991, the archive seemed to have been stripped, with the exception of the single report that Apsalikov had now stumbled on.

It was written in 1957 by scientists from Moscow's Institute of Biophysics and described the "expedition" by the special commission to Semipalatinsk. It was in Russian and marked "top secret," but Apsalikov's staff translated it for me. Its mixture of pedantry, vagueness, and occasional breathtaking revelation leaves me in little doubt that it is genuine.

The researchers reported finding widespread and persistent radioactive contamination across the towns and villages of eastern Kazakhstan. In mid-September 1956, radiation doses from breathing the air around Ust-Kamenogorsk reached 1.6 milliroentgen per hour, equivalent to roughly 140 millisieverts over a year. This was a hundred times the "permissible rate" at the time and appeared to be the result of "recent contamination," the report said.[3]

The following month, the expedition had moved on to take samples at a number of villages. The worst was Znamenka, halfway between the Polygon and Ust-Kamenogorsk. It would have been in the direct path of the fallout from the fateful test. "Near Znamenka, radioactive substances that affected the people and the environment fell out repeatedly for years," the report said. This fallout was "hazardous to health." The dangers were "more serious and dangerous than the district of Ust-Kamenogorsk." Military medical officers visiting the village had found three people with damage to their blood and nervous systems that they diagnosed as acute radiation sickness.

Kara-aul, a farming district southeast of the Polygon, was still contaminated by "hazardous" fallout from a test three years earlier, the expedition concluded. A test on August 12, 1953, had passed right over Kara-aul.[4] The investigators found more cases of apparent radiation sickness in the area, but the report said that medical staff found it hard to distinguish between the symptoms of radiation sickness and brucellosis, a disease caught through contact with local cattle. Researchers have since suggested that this "confusion" was a cynical ploy. Cynthia Werner, of Texas A&M University, pointed out that "Dispensary No. 3" was renamed the "Anti-Brucellosis Dispensary No. 4" in order to protect its secret mission of observing the impacts of radiation exposure on local people.[5]

People downwind of the Polygon were not just breathing in radiation on an alarming scale, they were also eating it. The 1957 report found "considerable radioactive contamination of soils, vegetable cover and food." Fecal samples taken from people at a collective farm just south of Ust-Kamenogorsk contained high levels of radioactivity, which stopped after families were given clean food imported from outside the area. The expedition called for a ban on eating local grain, "which was the most contaminated." It also suggested that it was "inexpedient to conduct atomic tests (especially ground explosions) before the full harvest from fields, and the shelter of grains and vegetables in storage, sheds, cellars etc." The dates of later tests show that this recommendation was not acted on.[6]

||||

The Soviet Union carried out more than a hundred atmospheric bomb tests in the Polygon between 1949 and 1962. Just four of the tests—in August 1949, September 1951, August 1953, and August 1956—delivered 95 percent of the dose of radiation to locals.[7] The first—which stunned the world by revealing that Uncle Joe had the bomb as well as Uncle

Sam—may have been the worst. It took place on the morning of August 29, 1949. The bomb's impact was maximized because it was detonated only 120 feet above the ground. The fireball hit the earth with huge force, lifting vast amounts of soil into the air, where it became radioactive and moved northeast in a narrow plume over dozens of Kazakh villages and across the border into Russia's Altai Mountains.

The fallout from the bomb made a direct hit on the village of Dolon, seventy miles downwind on the banks of the Irtysh River. Nobody was evacuated. "Because of official secrecy, they were not warned to stay in their houses. They were just left alone," Leonid Ilyin, director of Russia's Institute of Biophysics, told my colleague Rob Edwards more than forty years later, when the Soviet lid of secrecy was lifted.[8] Staying indoors would have done little good, however. Their wooden houses afforded scant protection from the radiation. Recent research using old Soviet radiation measurements estimates that the people of Dolon received an average dose of 1,300 millisieverts from this one test. That was enough to cause an epidemic of acute radiation sickness and likely many deaths. The citizens of Sarzhal, seventy-five miles southeast of the tests, received similar doses after a direct hit from fallout on August 12, 1953.[9] Ilyin says the twenty thousand inhabitants of Uglovsky, a rural area just over the border in Russia, got perhaps eight hundred millisieverts.[10] Over the years, these studies have concluded that some ten thousand people around the Polygon may have received doses of above seventy millisieverts, posing a potential risk to their health.[11]

One person who remembers those times well is Kaisha Atakhanova, who grew up in Karaganda, a city of half a million people 250 miles west of the Polygon. She later became a biologist researching frogs in radioactive ponds across the Polygon, and then an environmental activist. I interviewed her in London after she won the prestigious Goldman Prize for environmentalists. "People saw these huge mushrooms in the sky, but they didn't know what they were," she told me. After 1956, a few feeble efforts were made to protect people. "The military came to the villages and told people to come out into the streets, to lie down in ditches and cover themselves with white materials such as sheets or towels, and not to look up," she said. On other occasions, "they would give people red wine, which was supposed to be an antidote to radiation."

There were even evacuations "for a few days, until the dust settled. But the cattle and chickens stayed in the villages. Afterwards the people carried on living in their radioactive homes and with their radioactive animals." They even swam and caught fish in a lake created by a bomb test, where Atakhanova later researched the radioactive amphibians. "They didn't know it was dangerous."[12]

There has over the years been a lot of hyperbole about the health effects of the bomb tests at the Polygon. After the newly independent state of Kazakhstan shut the testing grounds in 1991, it renounced nuclear weapons and rebranded itself as a victim of the Soviet bomb program. For some years, a banner on its US mission website read "Nuclear Nightmare in Kazakhstan."[13] It encouraged reporters to visit a freak show of deformed fetuses preserved in formaldehyde in jars at Semey State Medical University and to tour villages where people were only too willing to tell their horror stories and bring out their sick children.[14]

How much of this should we believe? Everywhere has sick and deformed children; every hospital sees gruesome aborted or miscarried fetuses. The question is whether the area has more than might be expected, and whether anything can link that to radiation from the tests.

In the 1990s, Apsalikov's researchers at IRME began to assemble a database of those who had lived under the fallout clouds. The conclusions remain contested. Some analysts are certain they can see the impact. Marat Sandybayev, director of the Semey Oncology Center, says cancer rates downwind of the test area are twice those expected.[15] Another study reports more leukemia among children living less than 120 miles from the tests during the 1980s.[16] Some researchers are not always clear about whether confounding variables such as poor living conditions and bad nutrition might be to blame, rather than radiation.[17] Even so, the high doses of radiation received in the whole area around Semipalatinsk make such conclusions perfectly plausible.

||||

Today the Polygon is a magnet for those who enjoy "dark tourism." Guides head for the concrete towers that held the instruments measuring the intensity of the blasts. Radiation levels are still quite high close to these towers. Which raises the question of whether it is safe, even half a century on, for the natives living across the steppes of eastern Kazakhstan to catch the fish, graze their sheep, breathe the air, eat local bread and milk, or grow vegetables in their gardens.

Much of the radioactivity has decayed. Doses of radiation, even for those eating local produce, are orders of magnitude lower than in the era of tests. But some material lingers that sets Geiger counters clicking. Japanese researchers in 2006 found plutonium in soils around Dolon, the village that received the 1949 fallout cloud.[18] Nonetheless, Sergei Luka-shenko, deputy director of Kazakhstan's National Nuclear Center, who is responsible for managing the Polygon, is optimistic. He told me that, after some further cleanup, as much as 80 percent of the Polygon could be "returned to the community for regular economic use."

That still means he regards the remaining 20 percent—an area the size of Israel—as unsafe for the foreseeable future. As Apsalikov puts it, the testing zone "is not the Armageddon-like disaster that it sometimes is made out to be, but it is clear that some area will never return to nature, that the situation in others is uncertain and potentially dangerous, and the local people have experienced health risks as well as psychological stress."[19]

His mention of psychological stress is important. It is as much an impact of the tests as the radiation. It is frightening to think that you, or your children, are harboring disease because of bomb tests that took place on your land decades before. Likewise, not knowing what in your landscape is safe and what might kill you.

Such fear is corrosive. A local survey found that two-thirds of those alive during the tests believe that the fallout is responsible for bad health in their communities. They may be wrong. "Many local people seem to link problems of all kinds to the nuclear legacy," Apsalikov says. But whatever the radiological truth, "psychological stress and fear are an important and continuing legacy of the nuclear testing."[20]

Atakhanova agrees. She says well-meaning researchers often haven't helped in the way they have treated people seen as victims of the fallout. "Scientists came to test them and wrote scientific papers about them. But nobody helped them," she told me. "The people felt like guinea pigs."

But perhaps this passive victimhood, whether radiological or psychological, gets too much attention. Researchers who spend time living with communities in the Semipalatinsk badlands say there is a great deal of ambivalence about the risks. People often show a determination to somehow take ownership of their situation. Whatever their fears, many defy government orders and roam the radioactive steppes, growing fodder crops and picking wild strawberries. Tens of thousands of sheep and cattle are grazed close to the contaminating testing pads and water-filled craters.

The villagers have their own psychological survival strategy, says ethnologist Magdalena Stawkowski, of Stanford University. "Many claim to be 'mutants' adapted to radiation. They have come to accept their irradiated biology as genetically evolved and perfectly suited to their ecosystem. They believe themselves to be enhanced human beings who can survive in toxic environments." Such attitudes may seem odd to outsiders, she told me, but they may be better than always seeing yourself as a damaged victim.[21]

Chapter 6

Plutonium Mountain

Proliferation Paradise

In most places, atmospheric weapons tests ended with the Partial Test Ban Treaty of 1963. But, as in Nevada, the tests at Semipalatinsk did not end, they just went underground. They went to a granite summit in a remote part of the Soviet testing grounds, known to geographers as Degelen Mountain and to many in the nuclear fraternity as "plutonium mountain." The name arises because it is the only place in the world where you could go mining for the fissile metal at the heart of most of the world's nuclear weapons. The plutonium is not here because of natural geology. Plutonium is a man-made element. It is here because of almost three decades of experiments carried out by Soviet nuclear engineers in 181 tunnels they dug into the mountain between 1961 and 1989.[1]

Many of the experiments were conventional underground nuclear explosions that vaporized their fissile fuel. But Russian bomb makers had a particular penchant for testing the effect of conventional explosives on plutonium and for examining the behavior of plutonium in a variety of potential battlefield conditions. These experiments, often described as "minor trials," did not vaporize or disperse their plutonium. Instead, they left whole chunks of the stuff in the tunnels. Their environmental consequences, therefore, are far from minor.

Such experiments were not unique to the Soviet Union. In the early days, the US did much the same in the Nevada desert. But American researchers soon decided that the mess they were creating was too great and the loss of plutonium unacceptable. So early on, they switched to doing such experiments on a smaller scale, under controlled conditions, in the lab. But the Soviet nuclear scientists and weapons engineers had no such scruples. Degelen Mountain was their nuclear playground for almost three decades. When the Soviet Union collapsed in 1991 and they went home, they left behind inside those tunnels hundreds of pounds of weapons-grade plutonium: a treasure trove for anyone interested in finding radioactive material with which to make mischief.

Enter Siegfried Hecker, one of America's most celebrated weapons scientists and former director of the Los Alamos National Laboratory. In the early 1990s—when Boris Yeltsin was in the Kremlin and the former Soviet authorities were opening up to American technical advisors, financiers, and scientists—Hecker became the central figure in an extraordinary effort by the American nuclear establishment to pore over the Soviet nuclear legacy.[2]

Partly, he and his fellow investigators were curious. They wanted to check out what their adversaries had been up to for almost half a century of the Cold War. They also wanted to make Soviet nuclear hardware safe from accidents. But mainly they were growing increasingly worried at who else might be taking an interest in Russian fissile material. They knew that terrorists, rogue states, and simple criminals could find an Aladdin's cave of nuclear nasties across the vast territories of the former Soviet Union. The bad guys might also find ready assistance from cadres of government scientists and technicians, many of whom were not even being paid during the early 1990s.

Thanks to the breakdown of security across the former Soviet nuclear network of laboratories, testing grounds, military equipment, and power plants, "the situation in the Russian nuclear complex in the 1990s was the most dangerous in nuclear history," Hecker wrote in his book *Doomed to Cooperate*. The risks of outright nuclear war between the superpowers might have abated, but "the potential use of nuclear weapons somewhere in the world increased, because of the possibility of the theft or diversion of nuclear weapons or nuclear materials."[3]

In 1995, Hecker's nuclear odyssey through the Soviet system reached Kurchatov City, a city built to service the testing grounds of Semipalatinsk, and named after the country's nuclear pioneer, Igor Kurchatov. The city was once like Las Vegas without the hotels. It grew from nothing on the remote steppes of eastern Kazakhstan until, at its height, it had forty thousand inhabitants, living in the best conditions the Soviet system could provide.

By the time Hecker arrived, times had changed. Instead of a vibrant city full of the elite of the Soviet military-industrial complex, he found a near ghost town. It was more Mercury than Vegas. There were still a few thousand inhabitants, mostly servicing the Kazakh National Nuclear Center, a new agency awaiting a role. Even so, many suburbs were deserted, and the grand Palace of Culture sat abandoned. Curiously, the villa once occupied by Stalin's head of nuclear weapons production, Lavrentiy Beria, had been turned into a Russian Orthodox church.

"All I heard was a herd of horses running loose through the outskirts of town and an enormous number of ravens," Hecker wrote. To the south, across the Polygon, the steppes were similarly silent. The roads and towers, offices and weapons gantries created for hundreds of weapons tests were abandoned. What really horrified him, however, was Degelen Mountain.

He had heard that security was bad and that locals were digging into the mountain to find metal scrap, which had been turning up across the border in China. That was bad enough. Any of it could be contaminated with fragments of plutonium. Some prospectors might even be on the lookout for plutonium to sell. But the scale of activity took him by surprise. "I was expecting to see guys on camels, pulling out copper cables," he wrote in a report in 1998. Instead he saw "miles of trenches that could only have been dug by powerful excavating machines." The entire area was being mined for metal on an industrial scale.[4]

With jobs at the testing complex all gone, there was little else for locals to do but search the nuclear scrap yard for anything sellable. Since they had been hired to dig the tunnels in the first place, they knew well enough where it was buried. They even had a role model. The last Russian director of Kurchatov City had been fired in 1993 for doing his own looting.

Scavenging was widespread during the 1990s. The perpetrators used abandoned Soviet mining equipment and carried guns. None of the scavengers is known to have removed plutonium or to have suffered severe contamination as a result of their mining activities, but nobody can know for sure what was spirited away in those lawless years. Hecker estimated there might be 440 pounds of plutonium in the mountain. With the gates open and no guards anywhere to be seen, the material was "easily picked up, completely open to whomever wants to come."

Kazakh biologist turned environmentalist Kaisha Atakhanova told me in 2005 how villagers she knew would regularly dig up dumps where the departing Russians had buried military equipment. They found planes and tanks that had been left out during the weapons tests to see how they would be damaged. During her research into radioactive wildlife, "we went past these dumps regularly, and I saw that they were gradually getting smaller and smaller." The tunnels of Degelen Mountain were a favorite of the scavengers. "The Americans were helping to block the openings to mines where there had been underground tests, but the locals were reopening them to get at the valuable scrap inside."[5]

For more than a decade, and in near-total secrecy, Hecker fought endemic secrecy and mistrust to complete a $150 million project to make the plutonium mountain safe. He oversaw as local workers plugged tunnels, filled boreholes, and removed huge amounts of contaminated equipment. It was laborious and sometimes dangerous work.

A picnic on the site in October 2012 marked the completion of the cleanup. All was now safe, with all the tunnels concreted in and guards on duty, the assembly was told by Sergei Lukashenko, the new Kazakh boss in charge of keeping the mountain secure. He even had American military drones to help spot intruders. There was still plutonium in the mountain, but access to it is now "impossible," said Lukashenko.[6] That is reassuring. But plutonium has a half-life of thousands of years. How long will the guards and drones be around? How long will the concrete hold? Will local knowledge of the riches in the mountain outlast the diligence of the security?

Maybe not for long. Only months after the celebratory picnic, Harvard University's Belfer Center reported that a Kazakh survey team had found five more areas around the mountain where hitherto unknown plutonium experiments had taken place. They contained enough plutonium, at high enough concentrations, to pose a proliferation risk. The center quoted Byron Ristvet of the US government's Defense Threat Reduction Agency saying that "in some cases ... a guy with a pick-up truck and a shovel could accumulate enough [for a bomb]." The plutonium mountain, it seems, may not yet have yielded up all its secrets.[7]

In the West, nuclear authorities debate with environmentalists how to keep radioactive waste containing traces of plutonium safe during its long radioactive decay. How deep underground should it be buried? In what geology? What are the risks from earthquakes, rising sea levels, and changing climate? But in the plutonium mountain, lumps of the metal— even a shaving of which would kill, and a handful of which could make a bomb—sit around close to the surface, with no more protection than a cap of crumbling concrete. It is not a nice thought.

IIII

Things may not be much better on the old British bomb-testing grounds of South Australia. There were "dirty deeds" there, my Melbourne-based *New Scientist* colleague Ian Anderson reported in a 1993 exclusive. "Fresh Evidence Suggests That Britain Knew in the 1960s That Radioactivity at Its Former Nuclear Test Site in Australia Was Worse than First Thought. But It Did Not Tell the Australians," ran the headline. Certainly it did not tell the Maralinga Tjarutja, the Aboriginal people whose 1,300 square miles of territory had been used for the tests. But nor, it turned out, had it let on to the Australian government in Canberra.[8]

As a former dominion of the British Empire, Australia had been keen to help Britain develop its own nuclear weapons during the Cold War. It allowed the British to test atomic bombs first at the uninhabited Monte Bello islands, off Western Australia; then at Emu, in remote South Australia; and finally from 1957 at nearby Maralinga. Local Aborigines soon noticed the fallout. They called it *puyu*, meaning "black mist." Around forty-five Aborigines were enveloped by *puyu* near an Emu test in 1953. Some suffered skin burns, but no scientists seem to have checked out press reports that up to fifty may have died.[9] We still don't know their fate.

Besides detonating bombs, the British also carried out plutonium experiments similar to the Soviet experiments in Degelen Mountain. They were on a smaller scale, but were carried out al fresco, rather than in tunnels like the Soviets or in labs like the Americans. They released what Anderson called "jets of molten plutonium" across the land. Few people today have even heard of them. Even so, they were the cause of most of the problems highlighted by Anderson in his *New Scientist* scoop. For while bombs distributed their fallout widely, these "minor trials" left toxic remains in potentially lethal concentrations, often attached to military equipment left behind when the work ended. The fifteen "minor trials" conducted between 1961 and 1963, known as Vixen B tests, released approximately fifty pounds of plutonium around Maralinga. There was supposed to have been a cleanup afterward. Contaminated equipment was buried in twenty-one pits capped with concrete. The British weapons laboratory at Aldermaston said at least 90 percent of the plutonium was buried in the pits.[10]

But in 1984, when the government in Canberra wanted to hand the decontaminated land back to the Tjarutja community, they did a final check that found plutonium everywhere, mostly attached to fragments of contaminated equipment that had never made it to the pits. Peter Burns, of the Australian Radiation Laboratory, estimated there might be ten times more plutonium at the surface than the British had claimed—three million fragments in all: "People could just pick them up." Given the half-life of plutonium, a child playing in the contaminated dust anytime in the next few thousand years could inhale more than 460 millisieverts a year, according to a joint Australian, American, and British technical advisory committee set up in the wake of the Burns discovery.[11]

The findings were kept secret for several years, apparently while the Australian government persuaded the British to pay for another cleanup.[12] They were leaked to Anderson in 1993 once a deal was agreed.

The second cleanup began in 1995. Hundreds of thousands of tons of contaminated soil were scraped into burial trenches, and plutonium in the pits was reported to have been further immobilized using electric currents to convert the metal into a hard, glasslike rock, a process called vitrification. With $100 million spent, the potential dose to the hypothetical future child was now below five millisieverts per year, not much above global average background radiation levels.[13]

In 2014, at a ceremony in Maralinga, the final seven hundred square miles of the testing range was finally handed back to the Tjarutja community. And the rest of us can now take minibus tours. But even the latest cleanup was limited. About forty-five square miles is still circled by boundary markers that warn it contains "artefacts of the nuclear test era, including items contaminated at low levels with radioactivity."[14] Inside that area, a future child could receive sixty-five millisieverts in a year, claimed an angry Alan Parkinson, who chaired the original technical advisory committee that called for the second cleanup. It was, he said, "a cheap and nasty solution that wouldn't be adopted on white fellas' land." Some of that plutonium has a half-life of twenty-four thousand years. That, Parkinson acidly noted, is probably rather longer than the smart new warning signs will last.

Part Two

Cold War and Hot Particles

THE COLD WAR was a scary time. The arms race was in full swing and the risks of a nuclear conflagration were high. Everyone lived in fear of what was called mutually assured destruction, or MAD—satirized in the 1964 movie *Dr. Strangelove*. Behind the scenes, efforts to manufacture the plutonium needed for most of the world's atomic weapons at hastily constructed atomic cities reached a fever pitch. Accidents were bound to happen. And they did. In the space of four weeks in 1957, three massive fires and explosions ripped through the plutonium manufacturing and processing plants of the world's only three nuclear powers—the US, the Soviet Union, and Britain. The fallout from those accidents haunts the nuclear world to this day, nowhere more so than behind the Urals, at Joseph Stalin's answer to Hanford.

Chapter 7

Mayak

"Pressed for Time"
Behind the Urals

As the nuclear arms race began, the superpowers played hide-and-seek with their bomb-making factories. Just as America kept its nuclear plants outside the range of enemy bombers in the High Plains and beyond, so Joseph Stalin's brutal overseer of the Soviet bomb program, Lavrentiy Beria, hid his in atomic gulags behind the Urals. Construction began in 1946 in remote marshland an hour's drive from Chelyabinsk, the one big city in an empty region close to Russia's southern border. Conveniently, it was only about a thousand miles by rail across the steppes from the planned weapons testing ground at Semipalatinsk, in Kazakhstan.

Beria, like the British and subsequently most of the world, decided to make plutonium bombs. After Nagasaki, it was clear that, pound for pound, plutonium packed a bigger punch than uranium, and it was easier to produce. Beria's plutonium town was a copy of Hanford, with reactors turning uranium fuel into plutonium that could be extracted by chemical reprocessing. To confuse foreigners, it went by many names. Those have included Kyshtym, Base Ten, Chelyabinsk-40, Chelyabinsk-65, and today—still closed to outsiders but at least marked on maps—Ozersk. The complex itself was—and is—a company town, run by a state enterprise called the Mayak Production Association. By 1949, it had produced enough plutonium to make the first Soviet atomic weapon, which was detonated at Semipalatinsk in August of that year.

This had been done, however, at huge cost. Even by the standards of the day, the plutonium complex was built shambolically and with zero regard for the safety of its workers, their families, or indeed the landscape for hundreds of miles around. Accidents were frequent. Flasks containing lethal plutonium solutions regularly went missing. Conscripted soldiers were forced to clean up radioactive spills with rags and buckets.[1] Workers were routinely exposed to crazily high levels of radiation. There were at least seven criticality accidents, in which spilled plutonium briefly underwent a chain reaction, releasing a lethal rush of radioactivity.[2]

We know much of this because of the work of a Russian cardiologist named Mira Kosenko. She was recruited much later, in 1966, to track the medical consequences of the plutonium plant. This was secret work. She reported on the radiological carnage to her bosses at the Urals Research Center for Radiation Medicine (URCRM), but was not allowed to talk candidly to the victims she examined. She salved her conscience by spending years quietly gathering testimonies about past working conditions at the plant. Only after 1989 was she free to share her findings, first after visiting the Armed Forces Radiobiology Research Institute, a Pentagon "university" in Bethesda, Maryland, and later following a permanent move to live in California.[3]

Workers at the plant, she revealed, were almost routinely exposed to radiation doses that exceeded anything seen anywhere else in the world. This was well known to the authorities, but they dared not intervene because production targets had to be met. Conditions were especially bad in the shed where spent fuel emerging from reactors was "reprocessed" to release plutonium. There, some four hundred workers are known to have breathed in more plutonium than the worst known exposure among American workers.[4]

According to data collected by Angelina Guskova, a leading physician at the plant in the early days, three people died after exposure to 130,000 millisieverts in one accident. In day-to-day operations, worker Alexandr Aliev picked up 6,700 millisieverts in six months, after which his white-blood-cell count collapsed. He died of leukemia two years later. Another, Alexandrovich Karatygin, received ten thousand millisieverts. He survived the radiation sickness but both of his legs had to be amputated. "At that time, a man might receive 250 millisieverts in one working day," Guskova told Kosenko. She and other city doctors treated at least thirty workers who died of acute radiation poisoning, and dealt with more than 1,500 cases of chronic radiation poisoning, a debilitating condition rarely seen anywhere else in the world. There were five hundred cases in 1952 alone. Many workers died more lingering deaths. As of 2003, doctors had attributed 239 deaths to lung, liver, and bone cancers from plutonium.[5]

This human tragedy has become a gold mine for radiation scientists. Veteran American epidemiologist Ethel Gilbert, of the National Cancer Institute, who has studied Mayak's medical record in depth, says its victims provide "the only adequate human data for evaluating cancer risks from exposure to plutonium." Women, who made up about a quarter of the Mayak workforce and a higher proportion in plutonium-handling areas, suffered worst, she says.[6]

I was not allowed to visit Ozersk. It is a nuclear metropolis of one hundred thousand people but remains a closed city, even to most Russians. So I checked into a hotel in the regional center, Chelyabinsk, where I arranged to meet Sergey Romanov, the urbane, chess-playing director of the South Urals Biophysics Institute, which is based in Ozersk. The institute has monitored the health of the Mayak workforce since 1952. "Nowhere in the world has had exposure like this," he agreed. "Until 1955, even pregnant women worked on plutonium production." The women he called "the plutonium girls" were getting liver and bone cancers. Men were "dying by the hundred from lung cancers. . . . Leukemia rates were six times what they should have been." He said that hundreds of plutonium workers had died from a lung condition that doctors diagnosed as tuberculosis, but which was probably a condition known as plutonium pneumosclerosis, a growth of connective tissue in the lung caused by plutonium dust.[7] Those deaths should be added to the official toll.

The Mayak Production Association, the state-owned enterprise that ran the works from day one, clearly had a lot to answer for. The next person to visit me at my Chelyabinsk hotel was the company's longtime advisor on science and ecology, Yuri Mokrov. Surprisingly cheery and unabashed, he tried to explain to me what had gone wrong in the early days. "They were very pressed for time and lacked technical expertise. The area became very contaminated," he said. They were also, he added disarmingly, trying to make bombs using plant blueprints supplied by British atomic spy Klaus Fuchs. Their problem was that they didn't have the engineering know-how or safety expertise to go with the blueprints. They had to make that up as they went along. The mixture of ignorance and impatience was lethal.

Before having their plutonium extracted, uranium fuel rods from the reactor should have been kept in ponds for six months. That would have allowed the decay of the most dangerous radioactive isotopes, which mostly had half-lives measured in days or weeks. Instead, Mayak's managers ordered that the rods should go straight for reprocessing. "Some rods were dropped and shattered on the reactor floor," said Romanov. "They were highly radioactive. Yet workers took a shovel to remove the debris." Bad plant design meant that waste liquids from the reprocessing plant that contained "large amounts of short-lived radionuclides" routinely flowed down unshielded pipes near where many people worked before being stored in tanks.

It was these radioactive liquids, produced in huge quantities by plutonium reprocessing, that caused an explosion at the plant in 1957. That explosion generated the radioactive fallout that descended on the surrounding countryside, resulting in the creation of the world's first nuclear accident exclusion zone. I began this book with my visit to that zone sixty years later.

The liquids from reprocessing are among the most dangerous of all waste products from nuclear activities. They are extremely radioactive and that radioactivity generates a lot of heat. Moreover, back then nobody knew what to do with them. For want of anything better they had been accumulating in giant stainless-steel tanks not far from the reprocessing plant. By 1957, there were twenty of these tanks, each the size of a typical school classroom, lined up in a huge underground concrete bunker. The heat they generated was so great that they were kept from boiling only by a constant flow of water, circulating in the gaps between the tanks and the bunker walls.

One Sunday afternoon in September 1957, technicians noticed yellow smoke coming from the bunker. They suspected an electrical fault but found nothing. Before they decided what to do next, there was a huge explosion. "Tank 14 was completely destroyed," said Mokrov. The force of the explosion was so great that it blew the three-foot-thick concrete lid of the tank, which weighed 1,700 tons, about eighty feet into the air. "We estimate the explosion had a force of seventy tons of TNT."[8]

The problem hadn't been electrical. Instead, the flow of water through the cooling system of Tank 14 had been interrupted for some reason. The liquids in the tank had swiftly boiled, leaving behind sediment that exploded like gunpowder.[9] Most of the twenty million curies of radioactivity blasted into the air quickly rained down on the surrounding streets in a lethal black snow. The remaining two million curies formed a cloud that blew on gusty winds toward the open countryside northeast of the city, including the village I visited, Satlykovo.

There was no emergency plan, and nowhere for workers engulfed in the radioactive fallout to clean up. It was the weekend and the factory director, Mikhail Demyanovich, was eventually located at the circus in Moscow. Only in the early hours of the following morning did he order the evacuation of the area immediately around the tanks. But even as Mayak's scientists and senior plant operators left, around twenty thousand conscripted soldiers and prisoners from a local labor camp were brought in to clean up the widespread contamination.[10]

In the days that followed, these cleanup workers, known as liquidators, buried the radioactive debris in trenches across the city and dumped parts of the exploded Tank 14 into a nearby swamp.[11] This was lethal work. According to Mokrov, many liquidators received doses of 1,200 millisieverts, way over the official limit and highly likely to have caused radiation poisoning.[12] Yet, while Romanov's institute has tracked the health of permanent workers in Ozersk, most of the conscripted liquidators have never been followed up.[13] Their job done, they were shipped back to their gulags and barracks and forgotten about. Mayak's managers were under orders to get back to bomb production as quickly as possible. Within two months, there were new tanks and a restored cooling system. Plutonium reprocessing resumed.

Sixty years on, production continues. They wouldn't let me see, but according to Mokrov, the space once occupied by Tank 14 is filled with concrete. "The radioactive contamination and high dose rates of some areas within the plant site remain until now," he told me. "But we plan to decommission the complex before 2030." So that's all right then.

▌▐▌▐

The accident, often called the Kyshtym disaster, was at the time the world's worst nuclear accident. The fallout formed a long, thin plume that extended northeast of the plant for hundreds of miles. A few days later, managers at Mayak figured the villages nearest to the blast might be unsafe to live in. So, a week after the accident, troops moved into the three nearest downwind villages—Satlykovo, Berdyanish, and Galikayevo—and ordered their immediate evacuation. No questions asked: everybody out—and leave everything behind, including your clothes. "The advantage of the Soviet system was that we were able to take urgent action quickly," Mokrov told me with a wry smile. "Unfortunately, we couldn't inform the population about the reason for the resettlement. So they were scared."

People in villages a bit farther away from the explosion were initially allowed to stay in their homes but were told they could not eat that year's harvest from their fields and gardens, most of which had been ripening nicely when the cloud came over. The villagers were commandeered instead to dig up the produce and dump it all into pits.[14] In May of the following year, finding radiation levels still high, the managers decided to evacuate four more villages. A year after that, thousands more bemused people from another seventeen villages were led away. Finally, the men from Mayak fenced off an exclusion zone seventy miles long and six miles wide comprising the land with the worst fallout, which they called a nature reserve. They declared farmland as much as two hundred miles away to be unfit for use, or they deep-plowed it to bury the radioactivity.

There was no public announcement that there had been a disaster in the Urals. Not even the evacuees were told why they had to leave. Intelligence services in the West evidently got wind that something had happened. In 1958, a Danish newspaper reported rumors of a radioactive explosion in the Urals. It cited "diplomatic sources." Later, some Western geographers noticed that several villages had mysteriously disappeared from official Russian maps, while medical researchers began coming across reports in the Soviet scientific literature describing the results of "studies" into radiation effects that appeared to come from either an outrageously large experiment or, more probably, an actual accident.

Only in 1976 were the beans spilled, and then it was by accident. Zhores Medvedev, a biologist expelled from the Soviet Union four years before, wrote an article in my magazine, New Scientist, about the current state of Soviet science.[15] Toward the end, he made a passing reference to a "tragic catastrophe" in the Urals, during which an "explosion poured radioactive materials high into the sky." The story went around the world. Medvedev later said he had assumed the West knew all about it. He should perhaps have known that nuclear secrets were almost as tightly held in the West as behind the Iron Curtain.

Medvedev's description of what had happened was based on the recollections of liquidators and turned out to be wrong about quite a lot. He said the explosion was in 1958, not 1957, and that it was in a waste dump a couple of hundred miles from Ozersk. His suggestion of "hundreds dying" in the immediate aftermath may have been wide of the mark too. Such inaccuracies allowed Western nuclear chiefs to dismiss the story as "pure science fiction."

Eventually, in 1980, some American researchers successfully worked out roughly what had happened, based on reading between the lines of the Soviet scientific literature.[16] But Russian citizens were the last to know. They never heard a thing until after the collapse of the Soviet Union in 1991, when the files were opened. So, now they know the truth, and now that they don't live under the old Soviet regime, might the villagers be allowed to return to their land? Is it safe?

My guides from the Mayak Production Association brought a Geiger counter with us on our journey to Satlykovo. At around three millisieverts a year, radiation levels in the village air were barely above background levels. My guides encouraged me to wear a white protective suit, boots, and hat when we got out of the vehicle, but only to ward off local ticks that carry encephalitis. But radioactivity did lurk in the soil and vegetation. It was also in the lake across the overgrown fields from the abandoned village.

With the more virulent shorter-lived isotopes that were released during the accident now largely decayed away, the main danger lurking in the environment of Satlykovo was cesium-137. That has a half-life of around thirty years, so three-quarters had gone. "The only way to get a significant dose here now would be to eat large amounts of berries, mushrooms, or fish from the lake," said Oleg Tarasov, the young chief researcher of wildlife in the reserve, one of my guides. Such produce absorbs the surrounding radioactivity and imparts it in its flesh, he explained as we left the buried village and waded through the high grasses where the villagers' cattle once grazed.

The unnerving paradox of this village was that, however radioactive it might be, nature seemed to be have prospered in the absence of humans. The fenced-off reserve contained more than two hundred species of birds, said Tarasov. Among them were eagles and falcons. He had counted 455 species of plants, including locally rare species such as lady's slipper orchids. "The biodiversity is better here than in other reserves in the Urals," he said. "The animals understand they won't be hunted here, so they come. And they reproduce much better than elsewhere." The main exception seemed to be small mammals such as moles and voles that are closely connected to the ground. "They get much higher radiation doses and are more affected." For the rest, radiation was the least of their concerns.

The lurking radioactivity means that the ban on people returning is likely to remain for at least another hundred years, said Mokrov. The only people allowed past the gates to the reserve are scientists recording its wildlife and firefighters maintaining fire breaks to prevent blazes that could create a new radioactive cloud. Yet the scientists themselves seem largely unconcerned, even for their families.

A few hundred yards outside the gate to the radioactive reserve, one house remained in occupation. Our party had picnicked there before heading into the forbidden area. As we returned, Tarasov told me that the house had once been the home of a famous Russian ecologist, Gennady Romanov. It had been given to him as a base while he was doing research there after the accident. His widow still lived there, enjoying the peaceful countryside. She greeted us on our return. And it swiftly turned out that she was related to Tarasov. She was his mother-in-law. And Tarasov's family, including his young son, also called the house their home.

I was amazed. Why would he allow his own children to stay here, only yards from a zone that he himself declared too dangerous for villagers to return to? Tarasov was unabashed. "This place is quite safe in terms of radiation, despite its closeness to the exclusion zone," he insisted. "The fruit and vegetables that we grow in the garden do not contain an increased amount of radioactivity. Nor do the mushrooms and berries in the forest around the house. They can be eaten without any limitations. This has been verified by our laboratory." He had no fears for his son. Provided he remain outside the exclusion zone itself, he was safe.

As our party prepared to head back to Chelyabinsk, he said he planned to stay the night. He waved goodbye, cuddling the young boy for whom this is home. I wasn't sure whether to be reassured or appalled.

Chapter 8

Metlino

Even the Samovars Were Radioactive

The Techa River flowed quietly: spookily quietly. It was just a few feet wide. It meandered through rushes and broke into branches, forming small lakes as it flowed lazily downstream. Standing on the road bridge crossing the river, I could see several derelict buildings on its banks. The shape of a cluster of homes was reflected in the river. A little farther away along the road was a mosque; on the hill behind were the shells of disintegrated cattle barns. They were all that remained of the old village of Muslyumovo. The only sign of why the community had been abandoned was the rusting fences put up to keep people away from the river and off its floodplain. But they were long-since broken, and warnings signs had rusted until they were illegible.

Few people have heard of the Techa River. Certainly fewer than know about Chernobyl or Fukushima or Three Mile Island. Yet for many years half a century ago this nondescript Siberian stream was the world's most radioactive river. Probably more people were made sick by the effects of radioactive waste flowing down its channel than in all the other accidents put together. The story is all the more shocking because the disease and death happened slowly, over many years, with Soviet scientists secretly keeping track. And the source of the waste was the same Mayak plant discussed in the previous chapter.

In seven years from 1949, some 2.7 million curies of radioactive effluent was discharged down pipes into the river as Stalin rushed to build a nuclear arsenal to match that of the US. The lethal effluent flowed for hundreds of miles down the tiny river before being diluted by larger rivers that eventually flowed to the Arctic Ocean. More than one hundred thousand people who lived along its banks, drank its water, and grazed their animals on its marshy meadows were exposed to radiation at levels not seen anywhere else in the world before or since.[1]

The first inkling any of those villagers had that something was wrong came in July 1951. One day, "quasi-military types arrived at the villages along the River Techa. They pulled instruments from their small ponchos and . . . measured something, or took some kind of samples." That was how villagers later described the scene to Mira Kosenko, the medical researcher who analyzed their plight and collected their stories.[2]

Behind these comical cloak-and-dagger antics, scientists from the Mayak Production Association, the state enterprise that ran the bomb factory, were making the first measurements of radiation in what was by then one of the most radioactive places on Earth. Yet they were instructed to conceal the results of those measurements from the people who watched them come and go, unaware that their lives and landscape had been drenched with the complex's most dangerous waste products.

Back in the lab, the analysis of the samples of water, air, and vegetation shocked even the specialists. They found that a pond on the river in Metlino village, five miles downstream from the pipe that discharged the waste into the river, contained concentrations of the radioactive isotope strontium-90 that were hundreds of times the permitted level.

A secret committee was formed to decide what to do. Its chair was Anatoly Alexandrov, a nuclear physicist so eminent that the Russians later put his face on commemorative postage stamps. After due deliberation over several months, in early 1952, Alexandrov sent doctors to Metlino to report on radiation doses being received by its inhabitants and to judge their health. The doctors found that people who spent time in and around the Metlino pond were exposed to doses of a staggering fifty-four millisieverts an hour. As Kosenko put it, "women washing clothes or children sunbathing on the shore of the pond received in one hour the same exposure dose that, under current standards, is the annual limit for professional atomic industry workers and the lifetime limit for the public."[3] The doctors found that the average dose to teenagers in Metlino exceeded two thousand millisieverts in a year, with adults and young children receiving about half as much.[4]

It later emerged that up to a third of the radioactivity poured into the river in 1950 and 1951 ended up collecting in the water and sediment of the Metlino pond. Everything was grossly contaminated. The milk from cows grazing on the wet meadows along the river was radioactive; so were the vegetables grown in gardens irrigated by its flows, Kosenko said. "The samovars in which they boiled river water for tea . . . the bedroom sheets and underwear they washed in the river, the shoes they cleaned and floors that they mopped with river water, all were a source of radiation."[5]

It was no surprise that a third of the villagers—two hundred out of 578—had symptoms of chronic radiation poisoning, a condition seen before only among workers at the Mayak plant itself. Those symptoms included dramatically lower hemoglobin levels in their blood and a range of neurological and immune-system disorders. None of the symptoms was unique to the effects of radiation, but taken together they left little room for doubt.

Alexandrov's committee decided that Metlino should be evacuated, but once again nobody seemed in a rush. It took four years, until 1956, before the inhabitants all had new homes in a new Metlino village, built on higher ground away from the river. By then, almost two-thirds of them reportedly suffered from radiation poisoning. How many subsequently died we do not know. What we do know is that they died in ignorance. Through it all, nobody told the people of Metlino what had caused their sickness or why they were being moved. The official explanation was that the land was needed to expand a collective farm. Doctors who knew different were forced to repeat the story, even as they recorded the sickness and death among their patients.

How could such a calamity have happened? The reason is that, in their haste to arm the Soviet Union's first atomic bombs, Mayak's engineers had made no plans for the safe management of the huge volumes of highly radioactive waste emerging from their reactors and plutonium plants. So they adopted what they told themselves was a "temporary" solution— pouring most of it into the river that rose in a nearby marsh and wound past the plant. "Waste was just thrown into the Techa. We were in a race to build bombs; there was no time to do anything else," Sergey Romanov, of the South Urals Biophysics Institute, told me when I interviewed him in Chelyabinsk.

The plant managers knew from the start that their "temporary" solution was dumb. The Techa was tiny and within half a dozen miles of passing the plant it meandered slowly through an agricultural region where, for many people, it was their only source of water. Nonetheless, the plant poured into it an average of one Olympic swimming pool's worth of highly radioactive liquids every two hours. According to Kosenko, the plant's rulebook said the total radioactive discharge into the river should have been kept to ten curies a day or less. Yet at times the discharges reached one hundred thousand curies.

The best way to avoid admitting that huge disparity was simply not to take measurements. What the managers didn't measure, they didn't have to manage. And they had a ready excuse. Citing correspondence from the plant's head of external dosimetry, Kosenko wrote: "The secrecy surrounding production was used to explain why the discharge site had no flow meters or instruments to determine the radioactivity of the discharges."[6]

The worst of this unbelievable poisoning of the communities along the Techa came in April 1951. Heavy rains caused the radioactive river to overflow its banks, smearing lethal isotopes across its wide floodplain, including grasslands where villagers grazed their cattle, made hay, and grew vegetables. William Standring, of the Norwegian Radiation Protection Authority, who has reviewed the data, says the flood caused "severe contamination down the entire length of the Techa" until it entered the larger River Tobol 160 miles away. Dotted along its banks were thirty-nine villages with some twenty-seven thousand residents.[7] Most received their highest doses of radiation during the floods.[8]

It seems that it was concern about this flood that prompted the first cloak-and-dagger checks of radiation along the river that summer. After that, mobile teams regularly headed out to villages to find signs of radiation-related sickness. The monitoring was systematized when the Urals Research Center for Radiation Medicine (URCRM) was set up in Chelyabinsk in 1955. But even as the scale of the disaster became clear, the pace of efforts to protect villagers downstream of Metlino was glacial. It took until December 1952 for the first piped water to reach villages. After that, bans were imposed on use of the river, though "locals mostly ignored the restrictions, because nobody had told them why they were necessary," according to Louisa Korzhova, a retired nuclear physicist I met in Chelyabinsk, who had set up a nongovernmental organization (NGO) called Kyshtym 57 to represent the victims of Mayak. Perhaps as a result, new cases of chronic radiation syndrome in villages along the river did not peak until 1956, says the current director of the URCRM, Alexander Akleyev.[9]

In 1955, the authorities erected barbed-wire fences along the river's banks and finally started to move people out of another nineteen riverside villages. Homes were bulldozed, and all traces of the villages were removed from maps and official records. By 1960, with some eight thousand people moved and most of the villages empty, 940 cases of chronic radiation sickness had been diagnosed. Other casualties were emerging too. Researchers later concluded that around half of the ninety-nine cases of leukemia recorded among Techa riverside residents during this period were a result of radiation.[10]

Finally, after almost a decade of poisoning, it seemed as if all was now safe—except that one of the biggest and most contaminated riverside communities was somehow left in place. Muslyumovo, a village of two thousand people some twenty-five miles downstream of Mayak, remained inhabited until 2006.

After my interviews in Chelyabinsk, I drove first to see what remained of the old Muslyumovo, which I viewed from the road bridge crossing the river, and then to the new Muslyumovo, a couple of miles away on higher ground. It looked like any modern publicly built housing estate almost anywhere in the world. Its small but neat houses had red iron roofs and picket fences. There was a school, a clinic, and government offices. The residents' goats and cattle, having forsaken the river floodplain, now wandered the wide streets.

Many residents, keen to highlight their plight and how the authorities ignored them for so many years, tell how the old radioactive village briefly became a news story in the early 1990s, after a visit by President Boris Yeltsin. Local youths would charge journalists accompanying the president twenty dollars to photograph them swimming in the radioactive river. But still nobody came to rehouse them. They point out that after their belated move, researchers at the URCRM tested them and found that they had more strontium-90 in their bodies than residents of all the former river communities.

Some there say they suffer from diseases caused by years of exposure to the radioactive river. Nazhiya Akhmadeyeva gave birth to two sons while still living in the old village. One has a twisted spine and epilepsy that a government expert committee arbitrating compensation attributed to the effects of radiation. The other has hydrocephalus—"water on the brain"—that has left him severely mentally retarded.

Why the five-decade delay in evacuating Muslyumovo? There are many theories. The residents blame the fact that they are almost all Asiatic Bashkir people. They are looked down on by European Russians, they say. While "Russian" villages were evacuated, they were left as guinea pigs to see what would become of them. Mayak's Yuri Mokrov angrily dismissed that theory when I put it to him. Yet he was at a loss to provide a more convincing explanation. "Maybe the scientists testing the water came on a day when radiation levels were low," he suggested. I'm not sure he believed that. I certainly didn't.

IIIII

The story of the Mayak plutonium complex is testimony to the sacrifice of human life as Soviet bomb makers sought to catch up with the US in a deadly arms race. It is also testimony to the remarkable and chilling ability of Soviet medical researchers secretly to follow the fate of the victims. My last visit was to meet researchers at the URCRM, which has been tracking thousands of victims since it was set up in 1955. At its bare offices in a suburb of Chelyabinsk, I was ushered into a giant meeting room. I sat down with Ljudmila Krestinina, its current chief epidemiologist, successor to the estimable Mira Kosenko.

Since the late 1950s, Krestinina said, the URCRM has been monitoring two main groups. One comprised the thirty thousand people exposed to radioactivity along the banks of the Techa River. The other was made up of the twenty-two thousand exposed to the fallout from the 1957 explosion. In both cases, she said, doctors and the researchers from her institute had for many years been forced—under threat of imprisonment—to lie to their patients about why they were sick. Radiation sickness, like the plutonium plant itself, had been a state secret.

The researchers were otherwise assiduous, however. In the early days, they diagnosed chronic radiation sickness and reported it without fear. They worked hard to establish the doses suffered by villagers to see what they could learn about how much radiation caused sickness. If children drank milk from cows grazed on the river floodplain, the URCRM knew about it. If villages hunted waterfowl, that too was recorded. Since the end of the Soviet Union, they had made new assessments and checked the old data. "We have calculated the annual dose for each organ. So if we analyze, for instance, stomach cancers then we can compare the rates with doses to that organ," said Krestinina. She had recently been following the fate of more than twenty-four thousand children of the poisoned villagers, looking for any sign of genetic effects cascading down the generations. She had found none so far.

In recent years, she had collaborated with Western epidemiologists, who are keen to mine data they regard as unique. The doses of radiation experienced around the Mayak plant are much higher than those from any other large nuclear accident, including Chernobyl. And they are quite different from those received by Japanese bomb survivors, who were exposed to a single blast of radiation. The data could help resolve a number of debates, such as whether there is a threshold below which radiation does no damage; about the risks of thyroid cancer from radiation; and about whether exposure to radiation in the womb increases cancer risks late in life, Krestinina said. "After sixty years, our work is not done."

Some ask if we should be analyzing such data at all, given how it was obtained. During my visit it never occurred to me to question the current ethics of the researchers. But back in London, when I wrote up some initial findings, my editors at *New Scientist* magazine wrote an accompanying article arguing that there were "serious ethical issues about how the results were gathered and how health researchers should use them." The critical question was whether the people living along the Techa were, in effect, treated as guinea pigs in a radiological experiment. That might especially apply to the people in Muslyumovo. Such "research" using people as unwitting guinea pigs was rightly banned under international law, after the Nazi experiments on people in concentration camps. Consent is fundamental to ethical medical research. My editors conceded that "it is debatable whether failing to evacuate villagers quickly, and [then] covertly monitoring their health, amounts to experimenting on them. But there is no question that records were collected without informed consent."[11] But if it has been collected, surely the best response is to report the data, and expose what was done to the victims.

Kosenko, one of those who harvested the data for the URCRM and gathered the human stories that lay behind them, has now moved to the US. Under contract to the US Armed Forces Radiobiology Research Institute, she wrote the story of what had happened, beginning with terse academic papers and later in more human terms. Living in retirement in California, she is now free to vent her anger. In one report she asked:

"Did the designers and operations personnel know that there were dozens of populated areas along the banks of the Techa? They did. They knew that these villages did not have a single well, and they knew that the only source of potable water for the people living there was the river. They knew! In the lobbies at high-level meetings, I asked the same operators: 'Why didn't you at least dig wells in the villages?' Their answer: 'We hoped that it would all be diluted by water and would be insignificant.'"[12]

They may have hoped, but they were wrong. The truth is that no communities anywhere in the world suffered such prolonged exposure to lethal doses of radiation as those living along the Techa River between 1949 and 1956.

||||

Where do things stand now? While most people have been moved out of harm's way, the legacy of damaged human health lingers. So too does the environmental threat from Mayak's continued manufacture of plutonium and the huge volumes of radioactive waste within the barbed-wire fences of Ozersk. There is still some radioactive pollution going down the river. In 2006, Mayak's then chief executive, Vitaly Sadovnikov, lost his job after a Russian judge found that he had been allowing the discharge of concentrations of strontium-90 that "exceeded intervention levels." This had "created the threat of substantial damage to the health of the person and the environment." The judge found that, as a result, radioactivity in water from the Techa River exceeded legal limits at several villages that were never evacuated. At one, Krasnoisetskoe, river water was still used for drinking.[13]

I asked Mokrov about this. He was defensive; Sadovnikov had been his boss. "The radiation danger was grossly exaggerated by investigating bodies," Mokrov said. "No harmful effects were recorded." Maybe so, but his insistence that his boss had not been convicted was a bit disingenuous. Sadovnikov was indeed let off, but only because of an amnesty to mark the centenary of Russia's first national legislature.

What is true is that any recent discharges of radioactivity from the plant are tiny compared with the past. But there remains a potentially serious legacy of old discharges. Much of what went down the river decades ago collected in sediment on the river's bed and in its marshes, backwaters, and flooded pastures. There is a constant risk of this material being washed back into the river by floods and taken downstream. "The contamination is redistributing all the time," said the deputy head of environment in Chelyabinsk region, Svetlana Kostina, when I met her at the URCRM offices. Mokrov had told me that the broken fences I saw along the river were no longer necessary because the river was clean. But she said that was not true. Places that appear safe now may not remain so, she warned. The restrictions on access to the floodplain were being reassessed using new data, but many would have to remain "for generations to come."

"People do now understand that they are inhabiting a contaminated area, but they still swim in the river and have ducks and geese for food," says Galina Tryapitsyna, a researcher at URCRM. Some communities still consume contaminated water and contaminated milk from cattle that eat hay mowed from the floodplain. Even far downstream, anglers may be hooking contaminated fish. Eating fish soup is a particular risk, Tryapitsyna warns, because it is made with bones that concentrate the radioactivity.[14]

The biggest time bomb in this landscape lurks upstream within the boundaries of the closed city. Back in 1951, in the aftermath of their secret investigations in the Techa villages, Mayak's managers decided to divert some of the nastiest waste into a small bog close to the plant. The bog covered just 110 acres and did not drain anywhere, so the hope was it could become a permanent radioactive sump.[15]

But in 1967, a drought dried up what had become known as Lake Karachay. Strong winds whipped up its radioactive mud, which blew downwind, almost following the trail of the plume from the 1957 explosion at the waste tanks. About a million curies fell to Earth. Then the rains returned, and thankfully, the lake never dried out again. The most radioactive body of water in the world eventually accumulated more than 120 million curies—similar to the total releases from the Chernobyl accident.[16] Not many outsiders have ever seen it. One is Dick Shaw, from the British Geological Survey, who remembers being taken on a rare visit to the lake in 1998. "We were taken in a lead-lined vehicle, because the radiation was so great," he told me. Without that protection, radiation doses from the lake would quickly have been lethal to him and his fellow visitors.

In late 2015, with new technology greatly reducing factory discharges, engineers finally ended discharges to the lake and covered it with rock and concrete. This certainly reduced the radiation risks to people nearby, but the threat the covered-up lake poses to the environment has not gone away. Out of sight may mean out of mind. Beneath the concrete there are several feet of radioactive sludge, and beneath that another layer of concrete. Some experts doubt whether this crude containment around the most concentrated mass of radioactive material ever placed in the natural environment will hold for long.[17] When it fails, the mess will most likely find its way back into the Techa River.

Chapter 9

Rocky Flats

Plutonium in the Snake Pit

A barn owl burst from the tall prairie grasses. Elk skittered among cottonwood trees down near an old stagecoach halt on a trail heading west. A shrew crossed in front of us and hurtled into milkweed, where monarch butterflies fed. Somewhere amid the rare xeric grasses were coyotes, moose, mule deer, prairie dogs, a handful of endangered Preble's meadow jumping mice, and more than six hundred plant species. Occasionally, rumor had it, black bears and mountain lions dropped by. "Welcome," said David Lucas, of the US Fish and Wildlife Service, "to Rocky Flats Wildlife Refuge."

This oasis of prairie biodiversity covering four thousand acres outside Denver, Colorado, was undeniably beautiful on a bright summer's morning. Its only building was a barn on the historical Lindsay Ranch, which had occupied the land for almost a century until 1951. That was when the ranch was commandeered to provide a buffer zone around a nuclear facility. The land had been kept largely empty ever since. But Lucas hoped to change that. With the secret nuclear buildings now demolished, he was preparing to open the refuge for public access. Once the visitors' center was built and the planned twenty-five miles of trails complete, he was reckoning on 150,000 visitors a year.

I enjoyed my sneak preview of what Lucas had in store for future visitors. My host boasted that his domain was one of the best urban wildlife reserves in America, and I could not disagree. But I had some questions. Was it safe to allow the public in? What about the particles of plutonium in the soil, fallout from the many tons of plutonium that the facility once handled? Even if the lethal metal was now safely buried, wouldn't the prairie dogs, rabbits, and earthworms bring it to the surface before long? Would he bring his own children to mess about in the grass?

During the Cold War, Rocky Flats was a secret place, almost as secret as the Mayak plant behind the Urals in the Soviet Union, the previous stop on my world nuclear-legacy tour. And for the same reason. Locals were told that the Rocky Flats plant was making household chemicals. The real purpose of the place, over almost four decades, was to turn plutonium sent from Hanford into grapefruit-sized spheres known as "pits," after the hard core at the center of a fruit such as an apricot. The work was done first by Dow Chemical and then by Rockwell International, as government contractors. The plutonium pits they produced formed the explosive heart of each of the seventy thousand bombs in Uncle Sam's nuclear arsenal.

At its peak, 3,500 people worked at Rocky Flats, all sworn not to tell the world what was going on. They and the companies were paid huge bonuses for meeting production targets. The pressure to deliver was intense, and that often meant cutting corners. As a history produced for Congress after Rocky Flats closed in 1989 put it: "Production was the number one priority. Actions were taken that may have been at a higher risk to safety than acceptable by today's standards."[1]

Nowhere was that more true than in the plutonium-handling area of Building 771, a subterranean labyrinth surrounded by barbed wire and security guards. Workers called it the "hell hole" or sometimes the "snake pit" for its miles of meandering pipes containing venomous liquids. In Building 771, a largely female workforce molded and machined the silvery-gray metal inside glove boxes.[2] This was dangerous work. The metal was toxic, radioactive, and carcinogenic. Its dust could lodge in lungs and zap sensitive cells with radiation. Yet, as the congressional history reported, plutonium "routinely" ended up in floor dust or spilled onto walls and ceilings.[3] The plutonium could also spontaneously ignite in the air at room temperature. Small plutonium fires were frequent in the glove boxes. Usually they were swiftly extinguished, but not always.

At about 10 p.m. on Wednesday, September 11, 1957, plutonium shavings spontaneously ignited in an unattended glove box in Building 771. The heat-detecting alarm had been disabled because its constant soundings slowed down production. The glove boxes were made of a combustible plastic. So the fire spread unnoticed to the neighboring glove box—and the next and the next. The building's ventilation system fanned the flames and spread the smoke. By the time company fire crews arrived, plutonium had gone up the 150-foot works stack, where it over-whelmed filters intended to prevent any radioactive releases to the outside world.[4] Contrary to standing instructions, the firefighters sprayed water onto the blaze. There was an explosion, as gases from the blaze mixed with the water and the plutonium dust. The blast showered the fire crews with the stuff.[5] By morning, the fire was contained, but an unknown amount of plutonium had gone up the stack or into the plant's drains.[6] Much of it was scattered across the buffer zone around the plant and beyond. After Mayak, this was the second of the great plutonium accidents of late 1957.

The safety culture at Rocky Flats in the 1950s was lackadaisical, to say the least. That night was no different. John Hill, who monitored radiation at the plant, later told an oral history project that even though lethal smoke was billowing from the stack, "it was our feeling that things pretty well stayed inside the building."[7] So neither he nor anyone else took samples of the vegetation outside or checked the contents of the site's drains for plutonium or anything else.

Nor were there any second thoughts. "Immediately following the fire, plant officials' priorities were focused on restarting production, not analyz-ing the fire," stated a report on the affair from the Colorado state public health department.[8] It was back to work; business as usual.

Nobody knows for sure how much plutonium exited Building 771 that night. The potential was huge, however. Up to half a ton of the metal was handled in the building's glove boxes each month. But, despite its value and radioactivity, in-house audits never successfully kept track of it. In 1956, up to 220 pounds, a fifth of the output, went unaccounted for.[9] The best estimate is that around forty-five pounds of plutonium was involved in the 1957 fire, and about a pound may have spread outside the plant buildings. It could have been a lot more.[10]

Nobody warned the people of Denver about the fallout from the fire. Assiduous readers might have spotted a four-paragraph news agency story carried in a handful of newspapers. Quoting Atomic Energy Commission officials, it mentioned there was "possibly a light spread of radioactive con-tamination" but it would have been "extremely slight."[11] The report passed unremarked. Back in the 1950s, nobody worried too much about a "light spread" of radioactivity. Building 771 was cleaned up, patched up, and put back into pits production within ninety days. It was no safer than before. Even internal recommendations to reduce fire risks, such as installing glove boxes made out of something that couldn't catch fire, were not acted on.

So twelve years later, on Mother's Day 1969, when another fire spread after the spontaneous ignition of plutonium inside another glove box, the same set of failings produced the same chain of events. Flammable plastic was still in use; there was nobody close enough to notice when the glove box started to burn; firefighters broke the rules by using water; plutonium fumes spread through the ventilation ducts; and a stack filter failed.

This time, however, the web of secrecy around the plant also failed. It was the late 1960s. The antiwar movement was in full swing; Greenpeace was getting going in Vancouver. Citizens living near the plant now knew that what lurked behind the fences and beyond the buffer zone was a weapons plant. And they swiftly heard about the fire. Congressional hearings followed, during which the Atomic Energy Commission was forced to admit that "if the fire had been a little bigger . . . hundreds of square miles could [have been] involved in radiation exposure and cleanup at an astronomical cost."[12] Citizen investigators also discovered that, bad as the 1969 fire was, there had been an even bigger one twelve years before, with fallout that had spread all over the area.

The cat was out of the bag. But the pits kept being produced and the plutonium kept leaking. In many respects things got yet worse. Fire or no fire, the plant had nowhere to get rid of the plutonium waste from routine processes, which accumulated in ever-greater quantities at the site. It was often dumped in hastily dug landfills or sprayed onto grassland around the plant. More than five thousand drums of waste liquids contaminated with plutonium were stacked at an outdoor compound, known as pad 903. The drums began to corrode and leak into the soil. An internal memo reported that rabbits living on the site were heavily contaminated, especially in their hind feet. Some of that radioactivity got into local creeks and ended up in the mud on the bottom of local lakes and reservoirs.[13]

Years passed. The waste problem grew ever more acute. A plant worker, Jacque Brever, alleged that employees were ordered to incinerate plutonium waste late at night. This was illegal. She was pilloried for her claim, and claimed some coworkers tried to take revenge by spraying her with plutonium. But her allegations about incineration were confirmed by FBI agents using aerial infrared cameras. That led to a notorious FBI raid on the plant in 1989. After that—and with demand for plutonium pits declining with the end of the Cold War—the government closed the site.

Would heads roll? A federal grand jury sat for three years to hear testimony about evidence obtained from the FBI raid. Two days after the jury approved indictments, a deal was struck between the Justice Department and Rockwell. Rockwell pleaded guilty to some minor charges, but the FBI evidence and grand jury conclusions were sealed forever.[14] The whole sorry story formed the backdrop to a compelling memoir by a former neighbor of the plant, Kristen Iversen's *Full Body Burden: Growing Up in the Nuclear Shadow of Rocky Flats*.[15]

After the plant closed, the core plutonium-handling areas were declared a Superfund site, qualifying for a federal cleanup, which was completed in 2005. The cleanup removed six plutonium-processing buildings and some eight hundred other structures, as well as eighteen million cubic feet of radioactive waste, much of it in landfill sites round the complex.[16] The federal government called it "the largest and most successful environmental cleanup in history." The core area with the landfills and buried buildings nonetheless needed safeguarding, and it was put under the control of the Department of Energy. But the surrounding buffer zone was made into a wildlife reserve under the control of the Fish and Wildlife Service, which laid plans for opening it up to public use. I had come to find out if that was a good idea.

||||

Denver is growing quickly. Driving out of the city, I could see that the pastures that once separated the remote nuclear facility from the metropolitan area had been largely replaced by housing developments, highways, and park-and-rides. Rocky Flats now had a lot of neighbors. They needed to be kept safe, but they might also like to treat the wildlife refuge as a valued green space and opportunity to commune with nature. It was a selling point for the residents of the new community of Candelas that I saw being built right over the chain-link fence from the refuge. The developers' brochure wrote lovingly of the "magnificent sweep of mountain pastureland" provided by Rocky Flats.[17]

Lucas was all for that. "Rocky Flats is unique," he said as we bounced across its unplowed acres in his government-issue SUV. Thanks to the past military presence, the prairie ecosystem in the plant's buffer zone "contains one of the last stretches of tall xeric grasses in the Rocky Mountains Front Range," he said. With a few amenities it would be ready to open its gates. I could see the attraction. But I could also see, through the fence that surrounded the core zone, the eroding sites of landfills, the notorious pad 903 with its plutonium from drums stored there during the 1960s, and grassy areas where liquid wastes containing plutonium were sprayed. And I was aware too that the land over which we drove, and where children might soon be frolicking, was flecked with the plutonium that had gone up the stack. How much? Official estimates suggest maybe a pound. Is that dangerous? That is where the disputes begin.

||||

The next day, in the nearby university town of Boulder, I sat around a kitchen table with a dozen scientists—chemists, meteorologists, engineers, and hydrologists. Several had become opponents of the plant years before, after conducting research into the area for government agencies. One was Gale Biggs, a meteorologist appointed by Colorado's governor in the 1980s to assess air monitoring around Rocky Flats. He had found that plutonium particles that went up the stack from the plant's regular operations were so small that they eluded the stack filters and wafted into the air. Next to him sat Harvey Nichols, an earnest, British-born biologist from the University of Colorado. Back in 1975, he had obtained a federal grant to study the fate of those particles. He discovered that snow close to the plant was "hot." Snowflakes captured the tiny particles so well that just two days of snowfall could deposit about fourteen million particles on every acre of the site. "There must be tens of billions of particles in the soil today," he said.

That seemed to be confirmed by a study by Boulder radiochemist Ed Martell, who had pursued the health risks of radiation ever since witnessing the bomb tests at Bikini Atoll. In 1972, after the second fire at Rocky Flats, Martell showed that plutonium in the top inch of soil downwind of the plant reached up to four hundred times background levels.[18]

I had been introduced to the kitchen-table brain trust by LeRoy Moore, founder of the Rocky Mountain Peace and Justice Center. He had sat on a public committee in the 1980s that pushed for the cleanup of the complex and its buffer zone. The committee was told that to completely detoxify Rocky Flats would cost $37 billion, but Congress would sanction only $7 billion. The cut-price alternative demolished the larger buildings down to ground level and covered them with soil. The soil also hid the presence of twenty-five miles of underground tunnels and pipelines. "They were not even cleaned but just grouted and left. They may contain up to a ton of plutonium," he said. Government drawings of what remains underground broadly confirm his story.

What might happen to all this plutonium, which will remain a hazard for tens of thousands of years as it slowly decays? Moore feared the worst. Soils are living ecosystems, he said. "Prairie dogs and other critters will burrow down for several feet and bring plutonium to the surface," he said. If Lucas was allowed to open the refuge, "children will be exposed to plutonium. And people will start taking plutonium out into their communities on boots and cycle wheels. Why would we allow that?" If breathed in, the particles could be lethal.

The scientists were joined at our table by the federal agent who ran the FBI raid on Rocky Flats in 1989. His cop's handshake and steely gaze seemed out of keeping at a table of concerned liberal intellectuals, but Jon Lipsky was still seething about the way his high-profile investigation had been treated. "We got betrayed by the attorney general," Michael Norton, Lipsky told me. Not only that, the legal ruling that sealed the evidence he had collected about mismanagement of waste meant that it was unavailable either to the public or to those doing the subsequent cleanup.

All four men were adamant that the public should be prevented from setting foot in Lucas's beloved Rocky Flats reserve, however good the ecology. Their case for such caution rested in part on the concern that a single particle of plutonium could lodge in a lung and cause cancer by constantly irradiating a handful of cells. This "hot particle theory" is not accepted by government scientists, as I discuss in a later chapter. According to government regulations, there is an acceptable level of plutonium in soils. That level is 9.8 picocuries per gram of soil. The Rocky Flats refuge meets that level—well, almost. Lucas had told me that when the Environmental Protection Agency sampled soil in the refuge in 2006, it found an average of 3.2 picocuries downwind of the production areas. No sample recorded higher than eleven picocuries. In a letter to Lucas five years later, the EPA had concluded that "the lands comprising this refuge are suitable for unlimited use and unrestricted exposure." That was enough for Lucas. "It's safe," he had told me unequivocally as we strode through his prairie. No visitor should have any fears.

The Boulder crowd contested that conclusion. And they believe the plutonium is destined to spread well beyond the borders of the reserve. One reason is the inevitable fires on the wildlife reserve. Another is floods. On my tour with Lucas I saw earth slips that had left ugly gashes up to three hundred feet wide across a former landfill site that overlooked a creek running through the wildlife refuge. The Department of Energy's custodian of the core zone, Scott Surovchak, conceded that because of the local geology, "slumping is very common" at Rocky Flats after heavy rain. Nothing much could be done to prevent it. Only constant repairs, it seems, will prevent the landfills and contaminated buildings and pipework from being exposed.

David Abelson, director of the Rocky Flats Stewardship Council, a consulting body funded by the federal government, told the *Denver Post*: "It's easy to scare people about Rocky Flats . . . But the data don't bear out the fear."[19] He has a point, but the data are not so clear-cut. There is already plenty of plutonium beyond Rocky Flats fences.[20] In 2010, Moore found it in dust he collected from the crawl space under a home downwind of the plant.[21] And there could be health consequences. In 1981, Carl Johnson, the health director of Jefferson County, which includes Rocky Flats, found 24 percent more cancers among men who lived in areas downwind of the plant that had increased rates of "respirable" plutonium in the soil.[22]

That was a controversial conclusion. A later study for the Colorado Department of Public Health and Environment found similar results but blamed them on urbanization. Even so, the official reaction to Johnson's findings did not encourage confidence. He was told that his employers had "lost confidence" in him, and he was fired.

In 2016, a citizens' action group, Rocky Flats Downwinders, asked locals to come forward if they thought they might have suffered illness as a result of its operations. The Downwinders were the last people I met during my visit to Denver. Unlike the aging male Boulder scientists, the group was led by young professional women who had become concerned that their communities were living next to a half-forgotten time bomb. Tiffany Hansen grew up four miles from the plant. "The wind was blowing plutonium dust from the plant to our house. I played in that stuff; I breathed it; I cut my face rolling in it," she told me over coffee in Arvada, the suburb closest to Rocky Flats. Across the table was Alesya Casse, a legal analyst whose father worked on the cleanup at Rocky Flats. She feared for future children visiting the refuge: "How is it safe when a visitor has a rambunctious child who grabs a handful of grass and pulls it up by the root and throws it, or the child falls and scrapes their knee on a hot spot?"

The Downwinders say the government has done no proper research to assess health impacts from plutonium in their communities. So they launched their own survey. The initial results, published soon after my visit, did not reveal any pattern strong enough to indicate cause and effect. That was never likely. The 1,700 people who replied to requests for health histories are far from a random sample. Carol Jensen, a nursing professor at Denver's Metropolitan State University, who was the principal investigator, claimed to see "some concerning patterns." Some 40 percent of the cancers reported by the responders were categorized as rare, against a national average of 25 percent. "And thyroid cancer, nationally the ninth most common cancer, comes second in our research," she said.[23]

A likely explanation for the prevalence of the rare cancers is that Casse had said publicly she was looking for rare cancers. People with thyroid cancer might also be more likely to reply because they know it is often linked to radiation. Casse's own concerns about the plant arose after she was diagnosed with a thyroid condition and then heard Kristen Iversen, author of *Full Body Burden*, talking on the radio about having the condition. Hansen's father, who lived much of his life close to the site, also died because of a thyroid condition. Even so, the findings make a case for further study.

Lucas has roundly dismissed the cancer concerns of the Downwinders. "We need to get people out here on the refuge. Then the fears will evaporate," he told me. But that is just what worries his opponents. Forgetting about the plutonium is the worst thing that could happen, they say. It is surely reckless to be too sanguine about a Cold War site when we know well that risks were taken, corners cut, records not kept, and lives sacrificed in the name of national security. We know too that the government balked at the cost of a complete cleanup. Under such circumstances, it is not unreasonable to fear that sometime, somehow, somewhere down the Rocky Flats trails, picnicking school parties or hikers taking mud home on their boots could become victims of some dark secret from the past.

The standoff in this argument seems to me to speak not so much about the radiological hazards—which are real, though probably small under most circumstances—but about our inability to talk sense about radiation. One side denies there is an issue to discuss; the other often lacks a sense of proportion about the extent of the risk. At Rocky Flats, three decades after the bomb-making factory closed, we are left with a conundrum: is this a brilliant urban wildlife resource or a potentially dangerous radioactive legacy? The weird but inescapable truth is that it is both.

Chapter 10

Colorado Silos

*Uncle Sam's
Nuclear Heartland*

Colorado has a rich tradition of opposition to nuclear weaponry. Many of the Boulder crowd who fought early battles against Rocky Flats also opposed the government's deployment of weapons-carrying ballistic missiles in silos across the north of the state. Through them, I met the doyen of silo-busters, Bill Sulzman. I asked him to take me on a tour of Colorado's other nuclear landscape—the prairie lands of the north, stretching into Wyoming, that are dotted with silos still containing weapons targeted to obliterate cities on the other side of the world.

For Sulzman, the battle started in 1972, when the first Minuteman intercontinental ballistic missiles were being lowered beneath the High Plains of America. Back then he was a radical priest in his thirties, and the conservative politics of the Catholic hierarchy were getting him down. "The church had bailed out on the Vietnam War and it was bailing out on nuclear issues," he remembered as we headed north out of Denver. He began appearing in court, charged with various offenses connected with his protests against the missiles. "I subpoenaed the bishop." That didn't go down well. In the end, he said, "the hocus-pocus was not for me. I quit before they could fire me."

Some of Sulzman's best friends and coconspirators, members of groups such as Citizens for Peace in Space, stayed frocked. They included Carl Kabat, a minister who broke into silos dressed as a clown, and peacenik nuns. "The nuns were breaking into the silos too, doing rituals like pouring their blood," Sulzman said. "The resistance round here has mostly been from religious groups."

Soon we were in Weld County, in northern Colorado. Its 3,800 square miles are peppered with thirty missile silos, hiding down country roads and sprouting amid the winter wheat. Weld County's population is just 270,000, but such is its firepower that if it were a country, it would be a superpower, probably with a seat on the UN Security Council.

The silos seemed oddly small to me. Their enclosures are about the same size as those occupied by a cell-phone tower. Covered with a 120-ton concrete lid and surrounded by a fence with barbed wire, they are unmanned. Even the farmers who host them barely know they are there. As one landowner related in John LaForge's book documenting the silos, *Nuclear Heartland*, "After you've walked round a barrel of dynamite for 20 years, and it doesn't hurt you, you sort of don't think about it."[1] The surface footprint is deceptive, however. The silos beneath are more than 150 feet deep. The missiles they contain—wired to local hubs, then to the Air Force's Francis E. Warren base near Cheyenne, Wyoming, and finally to the president—pack a punch. A punch they can deliver within minutes to a location on the other side of the world that is already programmed into the onboard computer.

Just west of Greeley, off Highway 85, is the Weld County Missile Site Park, constructed in 1961 for Atlas missiles. A nine-man crew once worked twenty-four-hour shifts here. During the Cuban missile crisis the following year, the missiles were readied for launch. The silos were larger then because the missiles were stored horizontally and had to be elevated to vertical for firing. As new missile designs came along, this silo was deactivated in 1965. It had a later life as a fallout shelter. Today you can book a tour with the caretaker and stay on the camping ground. "Come and enjoy the peacefulness" says the park's website, without a hint of irony.[2]

In the distance, over the river near Platteville, we spotted another High Plains nuclear legacy. The Fort Saint Vrain store is a giant concrete edifice where spent fuel from old nuclear power stations is stored while the Department of Energy works out some permanent place for it to go. The facility contains more than 1,400 spent fuel elements, encased in blocks of graphite. It is designed to withstand earthquakes, tornado winds of up to 360 miles per hour, and flooding six feet deep. Also time. It will probably be several decades before the DOE finds the fuel a final resting place deep underground.

We were by now approaching the main silo fields. Sulzman was a college student when Margaret Laybourn, pushing two of her children in a baby carriage, put up a lone protest sign outside the first Atlas missile silo constructed at the Warren Air Force Base in the winter of 1958. The sign, written in shoe polish, quoted Pope John XXIII: "Mankind must put an end to war, or war will put an end to mankind."[3] In that moment, she started a protest movement that has continued out here on the plains for almost sixty years. "I haven't seen Margaret in twenty years," Sulzman said. "She had eleven children. She was passionately antinuclear. Her husband, Robert, had been a marine on cleanup duty at Nagasaki. I think that influenced her."

We turned off the main highway and drove down Country Road 115 to our first silo, code-named N-7. After half a century, it was part of the landscape. Just a couple of masts and a bit of concrete, surrounded by chain-link fence topped with barbed wire, and a bit of tasteful gravel. Inside the fence, it had the look of a neat war memorial, which in a way it was. At the gate we heard the hum of the silo air-conditioning mix with the howl of the wind.

N-7 is famous among pacifists in these parts. There was a big action against Minuteman III there in 1998, Sulzman remembered. It was on the anniversary of the dropping of the Hiroshima bomb. Songwriter Dan Sicken and Sachio Ko-Yin, a New Jersey bookseller and member of the War Resisters League, temporarily disarmed the silo by taking sledgehammers to the lid and buckling the steel tracks used to slide it away for missile firing. Then they sat down and waited to be arrested. They were sentenced to forty-one months and thirty months, respectively.

Priest Carl Kabat and Sulzman were back at N-7 two years later. Kabat wore his clown costume; he liked to call himself a Fool for Christ. "Carl climbed the fence; placed bread, wine, and a hammer on the silo; and prayed," Sulzman said. "We were both arrested. They dropped the case against me because I didn't get inside the fence. I couldn't climb it because I'd had a hip replaced the year before. But Carl got eighty-three days in Greeley prison. We still come back here every two or three years. I was roughed up here once."

The silos came thick and fast now; a new one came into view about every five miles. We dropped by at N-5 and then came to N-8, on Country Road 113 just west of New Raymer. "This is another shrine," said Sulzman, enthusiastic now. He got two banners from the back of the car and stood in front of the gate. He and his partner, psychiatrist Donna Johnson, posed for my camera. Donna held "Nuclear Bombs Are Immoral and Illegal." Bill held "Remove Nuclear Bombs from Colorado."

N-8 became famous after three middle-aged and extremely determined Dominican nuns came here in October 2002, in the runup to the US invasion of Iraq. Dressed in white overalls, Carol Gilbert, Ardeth Platte, and Jackie Hudson cut the fence and recited a liturgy, while smearing crosses on the silo lid in their own blood. They were prosecuted for "damaging government property with intent to obstruct the national defense." "Nuns v Nukes" ran the headlines the next day. In court, the judge was hostile. He refused to hear their testimony that the silos were illegal under international laws on nuclear disarmament or to hear their expert witnesses on the topic. Despite rallies organized outside by Sulzman and others, the three were convicted and sentenced to thirty-three, forty-one, and thirty months, respectively.

But they won friends. "The woman who lived in the ranch that owns the land here, Doris Williams, became chummy with the nuns," said Sulzman. "She went on a women's peace delegation to Moscow." I wanted to meet Williams, but that was a long time ago and she had moved on. At the ranch, there were a couple of pickup trucks, and a large Stars and Stripes fluttered in the breeze, but no one answered when we called.

The "Fool for Christ" climbed into N-8 himself in 2004 and was back at N-7 in 2009, hanging a banner, cutting the fence, and getting arrested for criminal mischief. He spent 137 days inside before his court appearance. The judge convicted him but set him free.

||||

America still has nuclear heartlands—landscapes where, every few miles, a missile is standing by ready to be launched on the orders of the president. Most Americans largely ignore this reality. It feels as if it is from another age. Certainly, many of those who still protest—waving banners, shouting through the wire fencing, and holding prayers in neat kitchens in Republican-voting suburbs—are showing signs of age. As he hoisted himself out of his car to raise a placard one more time in front of an unmanned silo, Sulzman admitted: "There are no young campaigners now. People forget."

The heritage industry is moving into the silo business. Out on Interstate 90 in the badlands of North Dakota there is a Minuteman Missile National

Historic Site. It has a visitor center and a launch facility that is "open daily, weather permitting." Visitors can "look down into the silo at a Minuteman missile in its historic state, protected by a glass viewing enclosure."

But we shouldn't be fooled. Fully armed silos are still all around. The president is still followed everywhere by a man with a briefcase containing a button that can launch missiles from any or all of them. Three missile fields have been taken out of commission, but three remain—in a crescent-shaped swarm west of Minot, in North Dakota; scattered across a large swathe of central Montana around Great Falls; and east of Cheyenne on the borders of Colorado, Wyoming, and Nebraska, where 150 Minuteman IIIs are deployed in as many silos, controlled from the Warren base near Cheyenne.[4]

The Minutemen carry 330-kiloton missiles—twenty times the size of Little Boy at Hiroshima—and are powered by solid fuel boosters. If accidentally ignited inside a silo, the fuel could damage the warhead itself. There would probably not be a nuclear detonation, but there could be a release of plutonium across the High Plains. Thankfully that has never happened. The closest call was in 1980, when there was a fire in a Titan missile silo fifty miles from Little Rock, Arkansas, where the Democratic Party convention was going on. It ignited the rocket fuel and sent the warhead flying, though it was swiftly recovered intact.[5] Another fire, in 1965 at an Atlas silo in Arkansas, left fifty-three dead.[6]

Some, like me, still remember those thirteen days back in 1962 when the Cuban missile crisis had the world on the brink of nuclear war. After Fidel Castro died in late 2016, eighty-nine-year-old Joe Andrew, in an interview on public radio, told of "driving nuclear warheads across Montana and fixing them to the top of waiting missiles" during the crisis in preparation for war. All the while he listened to the radio because "my thought was, as long as there are words being exchanged between Khrushchev and Kennedy, there's unlikely to be a nuclear strike." It was that scary.[7]

Back then the pace of the Cold War was hot. After the Soviet Union launched the *Sputnik* orbiting satellite in 1957, the US was suddenly playing catch-up. To fill the "missile gap," different designs of surface-to-air missiles were produced and replaced with staggering speed. The first Atlas missiles were scattered across the ranches of Wyoming, Nebraska, and northern Colorado beginning in 1959. Then came Titan, in 1962. No sooner were those missiles installed than, in the aftermath of the Cuban crisis, they began to be replaced by silos containing two hundred Minuteman I missiles. These carried hydrogen bomb warheads, each a hundred times more powerful than the Hiroshima bomb and able to target their destination using early computerized guidance systems. By 1965, Minuteman IIs were being deployed, and after 1971 came Minuteman IIIs, which can carry three warheads.[8]

Ronald Reagan's vaunted MX missiles, part of his planned "Star Wars" program, could carry ten warheads each. But they proved temperamental. In June 1988, just after first deployment, one of them collapsed in a silo on the plains of Wyoming, setting off an indicator in the control room that said "missile away." The operating rules said that on no account should anyone attempt to restore power to a missile in this state. It could trigger an accidental launch and explode the missile, filling the silo with radiation. Someone restored power anyway. That was troubling. With the MX suffering other problems in other places, deployment was reduced and the fleet was taken out of commission in 2005.

So we still have the Minutemen. Thanks to a lessening of international tension, they carry only one warhead these days instead of three. The firepower of the nuclear heartland is reduced, but it remains huge. Tax dollars are still being spent on new "deterrents." A few weeks after my journey across the heartland with Sulzman, the Air Force public affairs office e-mailed me with the news that "Air Force reaches first milestone in acquisition of new intercontinental ballistic missile." It had just got approval from the Pentagon for the "new phase of its acquisition effort . . . for the Ground-Based Strategic Deterrent program."

After four decades lurking in the silos, the Minuteman III missile fleet was about to be replaced, along with its 1960s command-and-control systems. Not stood down, but replaced. Deployment of its successors was due to begin "in the late 2020s," the Air Force assured me. In other news, I noticed that back in Washington the Government Accountability Office had questioned a bill for modernizing nuclear weapons that had risen to $300 billion. President Obama's 2009 pledge to "seek the peace and security of a world without nuclear weapons" seemed somewhat distant. Then Donald Trump was elected president, ordered a review of nuclear forces, and began saber-rattling over North Korea.

We drove down Country Road 123 to the small rodeo town of New Raymer looking for lunch. It was a Sunday: the community hall and the fire station, the church and the water treatment plant all sweltered in the heat, and the café was closed. Then we found N-1. This is the hub that controls and potentially releases the missiles from all ten of the N silos. It has two staff on duty at any time, with a small cafeteria, maintenance shed, and security barracks. "When we last came here, we got M16s pointed at us," said Sulzman brightly. The gate was open, however, so we drove in and parked in a bay. Nobody came out to meet us.

We were about to find a door to knock on, just to check if the nuclear front line was taking an afternoon snooze, when we spotted an armored Humvee with a machine gun on the roof following our route and pulling in. Had it been tagging us? "You can't park here, it's a military facility," said the young airman, getting out. "What kind of a facility?" I asked. "You'll have to ask USAF in Cheyenne about that," he said. Okay, that's cool, I thought. Pretend it's a secret, if you like. With the sign on the gate saying "Use of Deadly Force Authorized," I wasn't going to pursue the matter further.

We left, headed into Logan County and north to Peetz, a grain-elevator town of two hundred souls near the state line. Then we turned west, past J-3, for about ten miles. You could easily miss the silos here along a ridge dominated by three hundred giant turbines that make up the Peetz Table Wind Energy Center, one of the largest wind farms in the United States. But at the corner we made a sharp right and there was J-7.

"They had an accident here last year," said Sulzman. An automatic alert had sounded during a routine missile test. Staff responded, but while trying to fix whatever was wrong, they did more damage. There was a "mishap" that cost $1.8 million to repair, according to a statement dragged out of the Air Force by journalists. The Pentagon was briefed; Congress was briefed; but an independent commission set up by Defense Secretary Chuck Hagel that was conducting a review of nuclear forces, following other "mishaps," somehow was never alerted. All looked calm at J-7, however, as we got out the banners one more time. Was there still something to hide amid the turning blades of the Peetz wind farm?[9] "They won't tell the rest of us what the initial problem was, nor what they did to fix it that went so wrong," said Sulzman in mock bemusement. "All we know is the cost."

You have a sense that maybe these days the brightest minds are not focused on the nuclear weapons silos. Militarily, they may be at the front lines in the event of a nuclear war, but these days they are understaffed and underinvested backwaters. Newspapers report drug and morale problems among the men and women in charge.

I wanted to drop by the Warren Air Force Base, the giant home of the 90th Space Wing that dominates Cheyenne and which was in charge of all the silos I had been visiting. I wanted to see its Intercontinental Ballistic Missile and Heritage Museum. I could have asked about N-1, and maybe the advisability of having a nuclear arsenal in the hands of a demoralized air corps with drugs issues. Somehow, even with several weeks' notice, none of the base's three thousand servicemen and servicewomen could be found to show me around.

Instead I went for dinner with some peace activists in a suburb of Denver, hosted by former nun Mary Casper of Pax Christi, an international Catholic peace group. Around the table were another ex-nun, a Mennonite, a woman from the Church of the Brethren, and a former Catholic priest. Despite their advancing age most were active in their communities, working with the homeless, volunteering in care homes, teaching immigrant children, and running women's shelters. "Our actions are about morality," said Casper. "Militarism is rampant in our country. They are using our money to kill people." Innocently, I asked if any of them had been sentenced to jail. They smiled. They all had.

A couple of weeks after my visit, someone was arrested for splashing red paint on the doors of a weapons plant in Kansas City during a dawn raid. The culprit was none other than the old priest, eighty-two-year-old Carl Kabat, taking time out from his residence in a retirement settlement three hundred miles away. For once, the "Fool for Christ" was released from custody before noon. He would, he said, be back again next year.

Chapter 11

Broken Arrows

Dr. Strangelove and
the Radioactive Rabbits

No missiles have ever been released in anger from the silos of the High Plains. But since the dawn of the atomic age, some nuclear weapons have always been on patrol far from America's shore, in submarines or loaded into bomber planes. Sometimes they have had accidents. One befell the small, sleepy Spanish town of Palomares, some thirty miles north of Almeria, on the tourist coast of the Mediterranean Sea. Above the town on January 17, 1966, a B-52 bomber on patrol with four hydrogen bombs crashed into its refueling plane. Seven crew members died, three of the bombs hit the ground, and a fourth plunged into the sea. The detonators in two of the bombs exploded, scattering particles of plutonium over Palomares and surrounding fields. The deadly particles began blowing in the wind and spreading across the land. Many of them are still there.

In the three months after the accident, 1,700 US personnel pored over the fields and beaches around the town, measuring radiation and looking for bits of their bombs. A submarine secretly searched for, and reputedly found, the bomb lost at sea. On land, the cleanup crews excavated 1,500 tons of radioactive soil and vegetation, sealed it in drums, and shipped it to the US military's nuclear waste facility at Savannah River, in South Carolina. They buried other drums of soil in pits on site, along with a crop of tomatoes, and they put fences around one hundred acres of farmland where the bombs had created craters.

All the while, they denied there was any risk to the locals and even told reporters that no nuclear weapons were involved. Then they departed for home. One of the biggest spills of plutonium onto nonmilitary land was quietly forgotten about. The Pentagon never revealed how much plutonium had been in the bombs, though a likely figure is around six pounds. Nobody knows how much remains around Palomares, or how far it might have spread.

Some troops involved in the cleanup complained of illnesses afterward, but there was no systematic follow-up.[1] Few Spanish locals were ever checked for plutonium either. Those who were tested later complained that they were never told the results. Even so, at the time nobody made much fuss. Spain was still under the iron rule of its military dictator, General Francisco Franco. Moreover, Palomares was poor. There were only two cars, one phone, and no running water in the whole town, residents told a reporter later.[2] Besides, the tourist industry was just getting going along the coast. Palomares wanted a piece of the action, and advertising your plutonium wasn't a good sell.[3]

Only later did it become clear that Operation Moist Mop, as the Americans called the original cleanup, was botched. In 2010, European Union inspectors heard that locals had filled their homes with potentially radioactive chunks of accident debris as "souvenirs." They found that rabbit hunters had broken down the fences around the contaminated craters. Yet rabbit meat was never analyzed for contamination. Radioactive materials from the bombs still lurked. Some areas could give members of the public doses of five millisieverts a year or more. That was only twice the background level. But the inspectors still decided that the remaining plutonium-contaminated soil—up to two million cubic feet—should be either treated to remove the radioactive metal or taken to America for disposal.[4] In October 2015, US secretary of state John Kerry and Spain's foreign minister agreed to a three-year project costing $35 million to finish the cleanup and conduct regular blood tests on the people of Palomares. They might even check the rabbit meat.

||||

Back then, when fully armed bombers were constantly in the air, such "broken arrow" incidents were disturbingly frequent. They were also routinely hushed up. There was also growing concern that a mishap could not just result in a horrendous accident but could even trigger nuclear war. The film *Dr. Strangelove*, released in 1964, in which an unhinged American general dooms the world, captured the dark ironies of the historical moment nicely.

Two years after Palomares, another B-52 carrying four hydrogen bombs near Greenland caught fire. The crew bailed out and the bomber crashed onto sea ice near the Thule Air Base in the far north of the island. As happened in the skies over Palomares, conventional explosives in the bombs detonated, dispersing perhaps thirteen pounds of plutonium across the ice and into the ocean. This time the cleanup took place not on a Mediterranean holiday beach but in twenty-four-hour darkness, with temperatures down to minus-75 Fahrenheit.[5]

There were few bystanders, but the secrecy was intense. Thule was on the front line in the Cold War. The bombers were on alert in case their Soviet counterparts should fly over the Arctic Ocean and head for American cities. Until the accident, it had not been known publicly that the Danish government allowed nuclear weapons on Greenland, a supposedly self-governing province. That revelation caused outrage, and few were inclined to believe the later American claim that the cleanup had been successful.

The BBC revealed forty years later that, once all the bomb fragments had been collected from the ice, it became clear that one of the bombs had not been recovered. It had fallen through the ice, which then refroze. As at Palomares, the US Navy sent a submarine to find it. The Danish government was told little of this. "For discussion with Danes, this operation [the submarine search] should be referred to as a survey repeat survey of bottom under impact point," said a declassified Pentagon document uncovered by the BBC. After months, the search was abandoned.[6]

The US military still denies that there is a bomb on the seabed. Even so, official concern created by the two crashes in Spain and Greenland was so great that the Pentagon ended constant flying of fully armed bombers. Soon afterward, the United States and Soviet Union signed an agreement to notify each other in future of broken arrows of other weapons mishaps, in case they were misconstrued as a "first strike." Any future Dr. Strangeloves might then be thwarted.

Chapter 12

Windscale Fire

"A Cover-Up, Plain and Simple"

The third nuclear accident in late 1957 was at Windscale, the British plutonium plant on the remote coast of Cumbria, in northwest England. Windscale, known today as Sellafield, was the British equivalent of Hanford in the US and the Mayak complex at Ozersk in the Soviet Union. The existence of the plant—and the scientific failings that led to the fire—had their origins in the breakdown of relations between America and Britain at the close of the Second World War. America was determined to keep the secrets of how to make atomic weapons to itself. In 1946, Congress passed the McMahon Act, which cut off all scientific collaboration, including with its wartime ally, which had provided much of the basic physics and many of the physicists used in the Manhattan Project. The British scientists were sent home, told not to contact their American comrades—and even told to forget what they had learned while stateside.

Britain was understandably miffed. The new prime minister, Clement Attlee, secretly decided that "as a matter of urgency" Britain should independently develop its own bomb. The coterie of British bomb scientists sent back by Uncle Sam were brought into Whitehall and asked to start again from scratch, using the breakthroughs in atomic physics that they had taken to the US, and what they remembered from Los Alamos.[1]

Mathematician William Penney, who had been one of Robert Oppenheimer's top advisors at Los Alamos, was put in charge. He decided early on to concentrate on making a plutonium bomb of the kind dropped on Nagasaki. That required an atomic reactor to make the plutonium. So, in the autumn of 1947, construction started on two primitive reactors at Windscale—an old wartime munitions factory. While British citizens still endured food rationing, money was poured into the top-secret project.[2]

So far so good. The returning scientists' memories were good. The reactors, known as piles, were completed in good time, along with the reprocessing works needed to extract the plutonium from the uranium irradiated in the piles. Penney's piles delivered the first slug of plutonium in early 1952. It was only the size of a coin and the weight of a billiard ball, but two months later it had detonated the first British atomic bomb on a ship off Australia's Montebello Islands. Britain had become the third nuclear nation. By then, however, the US had developed its much bigger hydrogen bomb, which at first it called the "super-bomb." So Penney's team was given a new job: to make a British version.

The problem now was that most of the work on developing the hydrogen bomb had taken place after the departure of British scientists. Only Klaus Fuchs, the Russian spy still at work in the heart of the British nuclear establishment, had been involved. His role still appears to be sensitive. Fuchs's biographer Mike Rossiter stumbled on an old government file in the British National Archive called "Miscellaneous Super Bomb Notes by Klaus Fuchs." It contained twelve pages of calculations but it disappeared after he mentioned it on a blog site. Rossiter says that, thanks to British official secrecy, we know less about Fuchs's contribution to the British bombs than we do about the help he gave the Soviet Union.[3]

By 1957, the next British prime minister, Harold Macmillan, believed he was in a race against time to produce a functioning hydrogen bomb. This arose from both fear and ambition. He feared any international treaty banning tests might soon prevent Britain from testing such a weapon. But he also had high hopes that if he could test such a bomb successfully, then America would resume its technical cooperation over future bomb development. The critical date in his diary was in October of that year, when he had a summit meeting planned with President Dwight Eisenhower. Come hell or high water, he needed a successful hydrogen bomb test by then.

The bomb makers at the Atomic Weapons Establishment, in Aldermaston, west of London, were ready to go. But fueling the new bomb required lots of plutonium and tritium, which they planned to produce by radiating an alloy of lithium and magnesium in one of the piles. As Macmillan's deadline approached, there was intense pressure on the coterie of young scientists running the Windscale works to deliver. Through the summer of 1957, they went flat out.[4]

||||

Britain's two plutonium-producing piles were modeled on those built by the United States at Hanford. Each comprised a massive oven, with concrete walls six feet thick, into which were loaded thousands of cans of uranium fuel packed together close enough to start a nuclear chain reaction. The cans were assembled within a honeycomb of graphite, a form of crystallized carbon, which acted as a "moderator" to slow down the neutrons emitted by the fission reactions and control the speed of those reactions. Even so, the heat generated by the chain reactions was tremendous. Hanford's piles were kept cool with constantly circulating water, but the British decided air would be safer. So massive fans—rather like aircraft propellers—blasted air through the piles and up the four-hundred-foot chimneys.

The system was primitive, but it worked. But as demands for more and more plutonium and tritium grew through the summer of 1957, the scientists running the piles were in a quandary. They knew that the intense reactions in the pile caused the graphite that contained the uranium to gradually swell and hold energy—known as "Wigner energy," after the Hungarian physicist Eugene Wigner, who discovered it at Hanford. Wigner energy was dangerous; it could cause a reactor fire. So the scientists periodically reduced the swelling by shutting down the piles and allowing the core to heat up gradually with the cooling fans turned off. This technique for releasing the Wigner energy was potentially dangerous because too much heat could itself trigger a pile fire. And the longer the delay before each release, the more dangerous it was.[5]

To make matters worse, alterations to the pile to manufacture more tritium had created hot spots that were unknown because thermometers inside were in the wrong places. And the piles had no operating manual. With the Windscale scientists cut off from the expertise of their Hanford cousins, everything had to be done by trial and error. As a rule of thumb, they made Wigner energy releases every twenty thousand "megawatt-days," which meant every few months. But as the leaves began to fall from the trees along the Cumbrian coast that autumn, Macmillan's high diplomacy and Windscale's nuclear science were coming into dangerous conflict. The site technical committee in early September extended the gaps between Wigner releases to forty thousand megawatt-days.[6]

So it was only on Monday, October 7, just two weeks before the Eisenhower summit, that a long-delayed Wigner release from Pile No. 1 began. In the first hours of heating the pile, the release seemed to stall. So on the morning of the eighth, the physicist in charge decided to try to kick-start the release by adding more heat. He appears to have overdone it. Over the following twenty-four hours, temperatures inside the reactor began to soar. Nobody seems to have noticed. At some point a can containing uranium fuel burst, but an alarm that should have warned of a burst never sounded. More cans burst. Only after lunch on October 10 did anyone get wise to a fire in the pile. Air samplers on the roof of a neighboring building recorded radioactivity pouring out of the pile chimney. Soon there was smoke. The filters on top of the chimney had been overwhelmed.[7]

Deputy works manager Tom Tuohy climbed onto the roof of the pile. Peering inside, he was confronted by an inferno. About 120 fuel channels, containing around three tons of uranium, were ablaze. So was the graphite. Temperatures inside the pile exceeded 2,400 degrees Fahrenheit—compared with an intended maximum of four hundred degrees. The reactor was ablaze and there were no rules for what to do.[8]

Tuohy's first idea was to create a firebreak inside the reactor. He called men from the cinema at the workers' hostel in the neighboring dormitory town of Seascale and told them to pick up scaffolding poles from a nearby building site and ram hundreds of deformed and jammed fuel canisters out of their channels and into the water troughs at the back of the pile.[9] When that failed, Tuohy tried blasting more air through the pile to reduce temperatures, but the oxygen only fueled the blaze. As a last resort, with eleven tons of uranium on fire, he decided to douse the inferno with water. But even as the hoses were turned on, he feared that water sprayed onto molten graphite might trigger an explosion that would blow up the plant. As one longtime foreman on site, Cyril McManus, put it in an oral history of the plant: "If it had exploded, Cumberland would have been finished. It would have been like Chernobyl."[10] But it worked. By morning the fire was out.

Most of those involved in this hair-raising story are now dead. But one man whose life certainly hung in the balance, a young chemist named Morlais Harris, is still alive. He remembers that Tuohy had sent him onto the reactor roof to monitor temperature sensors, where he remained for twelve hours. "Later they said everyone apart from the firefighters had been cleared from the site before they put water into the reactor," he told me. "Actually I was on top of the reactor throughout. They had forgotten me. When they remembered and came to collect me at ten o'clock the next morning, the whole site was flooded with radioactive water that had poured out of the reactor."

If workers got forgotten as the accident unfolded, so did the general public. Despite radioactive smoke pouring from the chimney, and the real risk of a massive explosion, there were no announcements of an emergency, and no evacuations.

Some locals knew a bit, of course. Windscale's dormitory community at Seascale was not known in the media as "Britain's brainiest city" for nothing. Harris's father, a shop steward at the plant who was at home that day, put a Geiger counter onto his lawn and watched its needle go off the dial. Many workers on duty at the plant sent word to their families to get out fast. "It was all by word of mouth," Tom Jones, who was thirteen at the time, told the oral history project. He feared the worst because his father, a manager at the plant, had once told him, "If smoke ever comes out of those chimneys, you run as fast as you can."[11]

For those not in the know, however, life went on as usual. Harold Bolter, a former PR man turned company secretary at the plant who wrote an unauthorized history of the place, said, "Women went shopping, pushing young babies in prams and pushchairs as Pile No. 1 spewed radioactivity into the air."[12] Less than a mile from the blaze, pupils continued playing hockey in the grounds of Calder Girls' School. According to one of them, Jenny Jones, "when I went home that evening, my mum and dad knew something had happened . . . they'd noticed the village had gone very quiet. Then we got a phone call from one of the daily papers, asking if there was any panic in the village."[13]

||||

After the secrecy came the cover-up. With the story front-page news, the official line was that the smoke going up the chimney was harmless and in any case had blown out to sea. In fact, the smoke contained a range of radioactive isotopes of cesium, strontium, and iodine, though the total of around fifty thousand curies was less than 1 percent of the Chernobyl release thirty years later. Nonetheless, winds at the height where the smoke was ejected did not take the curies seaward. Instead, they blew the cloud southeast across England, triggering an alert across the English Channel in the Netherlands, before passing on to Germany and eventually Norway.

Windscale's managers knew this very well. And such was their concern that in the weeks afterward they secretly sent site foreman McManus around the country, as far afield as Devon, in southwest England, collecting samples of soil and vegetation to check the extent of fallout. As my magazine *New Scientist*, an enthusiastic supporter of nuclear power, put it in an editorial on October 17: "Public confidence has been severely shaken by what appeared to be attempts to minimise the gravity of what had taken place at Windscale, and even more by the extremely late hour at which any precautions to safeguard public health were put into effect . . . night calls by police two days after the first [radioactive] leaks occurred suggest an unfortunate belated awakening to the degree of contamination that might in fact be involved."[14]

What of Macmillan? By now, thanks to Windscale's efforts, he had his plutonium and his bomb test. But he was panic-stricken by the accident. It badly undermined his case that British nuclear technology was up to scratch. Eisenhower remained supportive of resuming scientific exchanges with Britain, but when Macmillan flew home from his summit to read a draft report on the Windscale fire from Penney, he found it so damning that he recalled every copy and sealed them away for thirty years.[15]

According to the minutes of a meeting of the British Atomic Energy Authority (AEA), which had overall control of Windscale, Penney had found "that this accident might well have been very much worse, and that a similar or worse accident might have occurred upon a number of occasions during the last few years. . . . Publication of the report would severely shake public confidence in the authority's competence [and] would provide ammunition to those in the United States who . . . oppose the necessary amendments of the McMahon Act."[16] To head off any suggestion of a cover-up, the government put out a heavily sanitized version of events that stressed nobody was at risk and placed the blame on "faults of judgment by the operating staff" during the Wigner release.[17]

There was no mention that the accident had been waiting to happen and could have been much worse. Nor, naturally, did this version mention that the underlying cause—the government's demands for ever more bomb-making materials for Macmillan—overrode operational safety procedures for Wigner releases. None of that emerged until the publication of the Penney Report by the Public Record Office thirty years later, on which much of the narrative in this chapter is based.[18]

But for Macmillan the secrecy and subterfuge worked. Within a few months, the first joint meeting of American and British weapons scientists took place. They have continued ever since. In British government circles Macmillan continues be revered as a master manipulator.

"He covered it up, plain and simple," Macmillan's grandson and biographer, Lord Stockton, told a BBC documentary on the fiftieth anniversary of the fire.[19] But it was par for the course in the nuclear industry then. The AEA supported the cover-up on the grounds that, as its minutes explained, to tell the truth about the fire would "shake public confidence" and "provide ammunition for those who had doubts about the development and future of nuclear power." Expediency was all.

||||

There is a strange postscript to the story of the Windscale fire. Through most of the 1980s, I was news editor at *New Scientist* magazine in central London. One day in 1983, a man showed up at our offices with a tale about how he had uncovered an unknown and highly radioactive ingredient in the cloud from the Windscale fire twenty-six years before. It sounded an unlikely story. We got cranks like that from time to time. Except that I knew the truth about the fire at Britain's plutonium factory remained an official secret, and it had taken years to establish that the cloud of fallout from the fire had passed over Britain rather than blowing out to sea.

So, sustained by a strong coffee, I invited John Urquhart, bearded Friends of the Earth activist, statistician, and librarian at the University of Newcastle-upon-Tyne, into the newsroom. Within an hour, I was convinced by his story. The radioactive cloud contained small amounts of a metal called polonium-210, which is so radioactive it glows blue in the dark. A few specks, if ingested, are enough to kill—as former Russian agent Alexander Litvinenko discovered when someone dropped polonium into his tea in a London hotel in 2006.

Polonium-210 was a vital part of early British nuclear weapons. It was used to provide the neutrons that kick-started the chain reactions that exploded the bomb. With a half-life of 140 days, it was hard to find in nature and impossible to store for long. So British bomb makers manufactured it to order by putting cartridges of an isotope of bismuth into side channels in the Windscale piles. This was going on in Pile No. 1 when the fire took hold. The cartridges had leaked and some of the polonium was blown up the chimney. It was measured in the cloud as it passed over Harwell, the nuclear laboratory in Oxfordshire. A naval base on the Dutch coast recorded a sharp polonium spike in the air three days after the peak of the fire.[20]

Scientists at Windscale knew about the release at the time. But they had a secret to keep—that British bombs relied on polonium-210. The Americans had long since stopped using polonium. As one source in the nuclear industry told me in 1983: "Britain just did not want the Americans to know how we were making our bombs."[21]

This cloak of secrecy failed only once. There was a brief reference in a presentation by a British scientist to a UN conference on atomic energy the year after the fire. At the end of a list of isotopes in the cloud, including cesium-137, strontium-89, strontium-90, and ruthenium-103, John Dunster added "together with polonium-210."[22] His presentation was never published in a scientific journal and was soon forgotten—until the university librarian and part-time atomic sleuth Urquhart uncovered the conference proceedings twenty-five years later.

It is not clear if anyone at the time of the fire calculated the potential health impact of dusting Britain with polonium. But Urquhart took up the challenge. Dunster gave figures for the ratio of polonium to iodine in filters that monitored the cloud. So with the total iodine release public knowledge, Urquhart was able to estimate the total polonium release at about 370 curies. Published data on the high radiological risks of polonium suggested that was enough to have killed a thousand people.

Before publishing Urquhart's amateur findings, I contacted the National Radiological Protection Board (NRPB), which had previously estimated the death toll from the cloud as it crossed Britain at thirteen. Its spokesman soon came back to confirm that Urquhart's detective work was broadly correct. Really it had no choice. For, by chance, the man who mentioned the polonium release at the 1958 conference, John Dunster, was by 1983 the NRPB's chairman. To be fair, they had come clean. The NRPB's subsequent reassessment came up with a polonium release of 240 curies. Citing a recent reassessment of the dangers of polonium-210 that Urquhart hadn't seen, it reckoned that maybe only a few dozen extra people had died, at most.[23]

Of course, the releases remained small compared with natural radiation, and the extra cancers were statistically unnoticed among the larger number of annual deaths across Britain. But they were real nonetheless. And whatever the death toll, as *New Scientist* wrote at the time, the real question raised by the affair was "whether, to protect the weapons programme, the scale of the accident at Windscale was covered up." It was a serious question, and one that to my knowledge nobody in an official position has yet answered. Perhaps because they know the answer has to be yes.

Three major nuclear accidents involving each of the three nuclear powers of the day, all within the space of a few weeks in late 1957, may perhaps be judged a coincidence. But there are underlying themes that have a lot in common, not least the unseemly rush to manufacture as many bombs as possible at almost any cost, and the callous lack of frankness with the public about what was going on. In the long run that approach has made the nuclear industry—civil as well as military—blind to its own failings and has so totally destroyed public confidence in all things nuclear that it begins to look like the first doomed industry of the Anthropocene.

Part Three

Atoms for Peace

THE SAME TECHNOLOGY that produced nuclear weapons offered the chance for nuclear power that its advocates said would be safe and too cheap to meter. But old, reckless habits died hard. "Atoms for peace" was a good slogan, but the nuclear power stations intended to deliver it were soon hit by their own litany of accidents: Three Mile Island, Chernobyl, and, most recently, Fukushima. The fences around stricken power plants were extended to embrace giant exclusion zones too dangerous for people to live. Or were they? My journey to these forbidden places finds wildlife resurgent in the nuclear badlands. Even humans who brave the radiation seem to be doing surprisingly well. Has the world been suffering from "radiophobia"?

Chapter 13

Three Mile Island

How Not to Run a Power Plant

Generating nuclear energy is much like the first phase of making a plutonium bomb. The same basic process is involved. Uranium fuel loaded into a reactor is bombarded with neutrons to trigger a controlled chain reaction that generates large amounts of heat and turns some of the uranium into plutonium. The difference is that when you are making a bomb, the plutonium is what you want and the heat is a waste product, whereas when you are making electricity, creating the heat is the main object, while the plutonium may or may not be a handy by-product. As we saw in chapter 2, a pair of emigré European scientists working in Britain, Hans von Halban and Lew Kowarski, had proposed this new source of electricity to the MAUD committee in 1940. Nobody took much notice because the urgent prize was making a bomb. But with the war won, it was a natural move for the bomb makers to turn civilian and make electricity.

The first reactors to do this for a national electricity grid were built at Windscale. Even as the two weapons piles were shut down after the 1957 fire, four new reactors were at work just yards away. Opened by Queen Elizabeth in 1956, they became known as the Calder Hall reactors. The heat they generated no longer went up the stack. Instead it heated carbon dioxide gas in pipes that ran through the reactor. The gas then went to a heat exchanger, where it boiled water to make steam for running turbines.

The plutonium didn't go to waste, however. The spent fuel from the reactors was shipped across the site to the reprocessing plant, which extracted the plutonium to continue the production of British bombs.

The Calder Hall reactors became the prototypes for a fleet of British civilian power-generating reactors. Magnox reactors, named after the magnesium oxide cans that contained their fuel, sprouted across the remote coastal headlands of England, Scotland, and Wales in the 1960s and 1970s. None were close to cities, a decision driven by continuing concerns about their safety following the Windscale fire.

Other countries swiftly followed with their own designs of power-generating reactors. America's first, the Shippingport station, on the Ohio River in Pennsylvania, was opened by President Eisenhower in 1958. It was the first reactor with no ancillary military purpose. Eisenhower declared it part of America's "atoms for peace" program. Another hundred followed. Most were of a new design. While British Magnox reactors followed closely the design of the original wartime Hanford reactors, America mostly chose the pressurized water reactor.

The PWR was a scaled-up version of a military reactor first developed to power submarines. Instead of gas, it used pressurized water to cool the reactor and take its heat to the heat exchangers. First reconfigured for electricity generation by Westinghouse, the PWR remains the main nuclear workhorse around the world to this day. (Since Britain closed its last Magnox in 2015, the only survivor of that design is in North Korea.)

||||

These new civilian reactors were, many believed, going to usher in a new nuclear age. "Atoms for peace" would transform how we got our electricity, just as surely as the atomic bomb seemed to have transformed warfare. Nuclear engineers became cheerleaders for hugely inflated public expectations about the growth of nuclear-power generation. In Britain, Prime Minister Harold Wilson said that through nuclear leadership, his country could pioneer what he famously called the "white heat of the technological revolution." The American nuclear chief Lewis Strauss boasted that nuclear energy would soon be "too cheap to meter." He promised the country would have a thousand nuclear reactors humming away by the end of the century, rather than the one hundred actually built. There would, some said, soon be nuclear planes and nuclear cars.[1]

But the military culture of secrecy and cover-up at nuclear installations persisted. It was a culture that kept things handily quiet in the good times, with few questions asked about the costs or risk of the new technology. But it also meant that nuclear operators often failed to learn from their mistakes, covering up and excusing them even among themselves. And it proved catastrophic in bad times, when all public trust was lost. The accident that changed the way the world viewed nuclear energy came in March 1979 at Three Mile Island. Famously, it was an accident of the type forecast in the movie *The China Syndrome*, starring Jane Fonda, which had opened a week or so before.

Three Mile Island is a thin strip of land in the middle of the Susquehanna River in Pennsylvania. A PWR had been in operation there for only a year when it lost its cool. In the early hours of March 28, the pumps feeding heated water from the plant's "unit 2" reactor to the heat exchangers stopped working. Pressure in the cooling system's plumbing started to rise. A valve operated correctly to release the pressure but failed to close when it should have. Because a warning light in the control room malfunctioned, operators were unaware for eighty minutes that the valve remained open. During that time the cooling water continued to vent from the system.

Operators had shut the reactor down at the start of the incident, but fission products inside the reactor continued to produce heat. So without a functioning cooling system, temperatures in the reactor rose. The cans containing the reactor fuel began to burst, adding to the heat. A metal in the fuel cladding called zirconium reacted with the steam to produce hydrogen, raising the risk of a hydrogen explosion that could have scattered huge amounts of radioactive material across a wide area of Pennsylvania.

The problems escalated. Some of the coolant water was highly radioactive. There was a small release into the air, which rained down on the surrounding town of Middleton. One official said publicly that there could be a meltdown of the reactor, the so-called "China syndrome" of the movie. Even in a predigital age, this was now worldwide news. For five days the efforts to prevent meltdown or a hydrogen explosion dominated headlines.

In the end, neither happened. Releases of radiation, at around fifteen curies, were minimal—less than 1 percent of the release during the Windscale fire and a ten-millionth of the later Chernobyl release. The average dose to anyone in the vicinity was about 0.08 millisieverts—roughly the equivalent of a chest X-ray. Nobody is thought to have received more than one millisievert, a third of the annual dose from background radiation. No subsequent studies found any convincing evidence of health effects, though some tried.[2]

The accident had been the most serious to date at a civilian nuclear power plant. It left behind a big mess but a disaster had been averted. Ultimately, safety systems worked. Few outside the industry were satisfied, however. The media story of Three Mile Island became not about a community saved, but about bungled operations in the operating room: about confusion, the failure of backup systems, and botched safety procedures that suggested an industry operating not so much with technocratic efficiency as on a wing and a prayer.

The subsequent official investigation by a presidential commission, known by the name of its chair, John Kemeny, confirmed that. The Kemeny Commission concluded that the myriad failures as the accident escalated were mostly due to human errors—errors that stretched all the way from the Nuclear Regulatory Commission to the manufacturers and the chaotic management of the control room. Secrecy contributed. It emerged, for instance, that similar relief valves at similar plants had often failed in the past but that the information had never been shared.[3]

This culture of secrecy revealed by the commission reinforced a sense of institutional incompetence at the plant and within the industry as a whole. It meant that Three Mile Island's neighbors knew little of what happened inside a power plant, so when things started to go wrong they could believe the worst. With official information scarce and contradictory, fear stalked the communities near the plant. Altogether some 140,000 people fled their homes, and even official advice at one stage asked children and pregnant mothers to leave.[4]

The reactor was permanently shut down after the accident. Cleaning up the debris and making it safe took fourteen years and cost a billion dollars. The shell still awaits final dismantling. But it is the wider results that have been the longest lasting. The accident, as the British newspaper the Observer headlined it, caused a "Nuclear Loss of Innocence." Public fear and suspicion about all things nuclear grew sharply in the years after Three Mile Island, especially in the US. In the next five years, fifty-one orders for new nuclear reactors were canceled. New contracts entirely dried up.

On the day the accident began, the US had 140 operating nuclear reactors, with ninety-two under construction and twenty-eight more awaiting official approval. Afterward, all orders for new plants were halted and, at the time of writing, no new plants had begun generating power in America in the thirty-eight years since the accident.

||||

For the British nuclear industry, the tipping point that turned the public against nuclear power came four years later. It was not a power-plant accident, but a scandal at the reprocessing plant at Windscale, which extracted plutonium from spent reactor fuel. The British no longer needed plutonium to make new bombs. They had enough of those. The plutonium was potentially a major waste problem. But they had another idea: to turn the fissile plutonium into fuel for future generations of nuclear power plants. Such plants, known as fast breeders, were then under development at Dounreay, a research site on the north coast of Scotland.

The plan developed during the 1970s, when nuclear optimism was at its height. It sought to make Britain a world leader in fast-breeder technology and win it a global market in the new fuel. Nuclear advocates talked of a coming "plutonium economy." Spent reactor fuel was not a dangerous waste, they said, but a valuable resource. Plutonium for peace, you might say. The gung-ho nuclear engineers at Windscale persuaded the British government to expand reprocessing capacity by building a giant new reprocessing plant. That plant would take spent fuel not just from Britain but from across the world. But while enthusiasts saw this as a technological utopia, others saw things differently. Rumblings of discontent were growing at what seemed an industry out of control and able to tap government money at will. The scheme prompted the famous 1975 *Daily Mirror* headline that Windscale had concocted a "Plan to Make Britain World's Nuclear Dustbin."

With those words ringing in their ears, British nuclear bosses grew wary of bad PR. After Three Mile Island they decided to detoxify the Windscale "brand" by renaming it Sellafield, after the small local community obliterated to make way for the works. But only weeks after the launch of the new name, on Friday, November 11, 1983, alarms sounded across the Windscale plutonium complex, plunging the facility into its most serious crisis since the fire twenty-six years before.

During a maintenance shutdown to clean out the Magnox reprocessing plant, a shift worker had misread a penciled note on his operating instructions. Faced with several different liquid waste products in different pipes leading from the reprocessing plant, he mistakenly poured half a ton of highly radioactive solvent into two tanks intended to take much less radioactive waste liquids for discharge down the pipeline into the Irish Sea. There was no easy way of backing up, so into the sea that night went 4,500 curies of radioactive material, more than four times the normal officially recorded discharge down the pipe for an entire year.

The pipe stretches a couple of miles into the Irish Sea. But Sellafield's luck was out. First, the radioactive liquids entered an unusually calm sea. So instead of dispersing on the waves, a radioactive slick began to float back to the shore. Second, there was Greenpeace. The environmental group, which had long wanted to force the closure of Sellafield, had been surreptitiously taking samples of seawater at the end of the pipeline. They were not at their post when the "Saturday night special" went down the pipe. But on the following Monday morning, they found themselves in the middle of something rather nastier than they were used to. Their Geiger counters chattered madly. Greenpeace being Greenpeace, it took only a few hours for the world to know what they had discovered.

The slick stuck around. Harold Bolter, Sellafield's PR man at the time, related years later in a history of the plant that "bits of radioactive flotsam kept turning up on the beach, some of them giving off fairly high levels of radiation, certainly much higher than would be allowed on a public beach." This was no technical breach of safety rules. This was serious. Yet plants managers repeatedly misled him about the incident, he said. "I got the distinct impression that there had been similar discharges of solvent and crud in the past, but the strong tides and heavy swell of the Irish Sea could normally be relied on to carry the material away . . . I am convinced that if the slick had been dispersed in the sea as expected, Sellafield would have kept quiet about it."

The upshot was that, after several days of prevarication by the company, some ten miles of beaches along the West Cumbrian shoreline was closed for six months. The main concern was that, as Bolter put it, "children might pick up one of the pieces of radioactive debris and hold it for several hours, burning their skin. Worse still, a child might put such material—bits of plastic, rubber, or string—in his or her mouth and swallow it."[5]

The eventual court conviction for British Nuclear Fuels Ltd. (BNFL) led to a derisory fine of just $13,000, but the reputational damage was huge. "The assault by Greenpeace marked a turning point," says Andrew Blowers of the Open University, a former government advisor on radio-

active waste management, adding that 1983 "can be identified as the piv-
otal point [in Britain] when nuclear issues achieved widespread public
awareness." Sellafield, even more than Windscale before it, had become
"a byword for the dirty end of a dangerous industry."[6]

IIIII

The scandal intensified soon after, when a TV program, *Windscale: The
Nuclear Laundry*, found a cluster of leukemia cases among children living near the plant. Alan Postlethwaite, the local vicar, later told an oral
history of the Sellafield plant that "within quite a short period of time, I
conducted funerals of three children who died of leukemia." Statisticians
had told him to "expect one in twenty years, and we'd had three in twelve
months. . . . That put the frighteners on us." Jenny Jones, who had once
played hockey at the local school while the Windscale plant burned over
the fence, said that two children in her son's school class had the disease.
Whenever her own child got sick, she feared the worst.[7]

Medical researchers eventually concluded that the local leukemia rate
was many times the national average. Moreover, men who received a
cumulative radiation dose of more than one hundred millisieverts before
conceiving a child were likely to have suffered mutations in their sperm
that could have increased their chance of fathering a child with leukemia
by six- to eightfold.[8] Dozens of Sellafield workers fell into that category.
The report caused pandemonium, especially when one of the company's
doctors told journalists they were thinking of advising some workers not
to have families. The company changed tack and blamed the cluster on
outside workers bringing a mystery virus to a previously isolated coastal
community. But this was undermined by a later study paid for by the company, which found that viruses could not be to blame.[9]

Sellafield's medical staff had clearly been worried for many years about
what was going on there, especially the spread of plutonium around the
plant. They had arranged with local pathologists to secretly collect the
corpses of retired former workers and check them for plutonium. I was
partly responsible for uncovering this surveillance, which had echoes of
the secret monitoring of people downstream of the Mayak plutonium
plant in the Soviet Union.

In 1986, I was leafing through the internal bulletin of the government's
National Radiological Protection Board. I found an article by its medical

researcher Don Popplewell, describing research showing that plutonium levels in the lungs and lymph nodes of former Sellafield workers were hundreds of times higher than in the general population. I called Popplewell. He told me that Sellafield's chief medical officer, Geoffrey Schofield, who had died the previous year, had analyzed more than fifty corpses of former workers. There could be no doubt, he said, that Sellafield was the cause of the plutonium, though the health consequences were unclear. All this was done even though, as I wrote in *New Scientist*, it was "strictly illegal to examine autopsy tissue except to ascertain the cause of death."[10]

My article sank without trace. It was only twenty years later—after an unconnected scandal over illegal autopsies conducted on the corpses of children at the Alder Hey hospital in Liverpool—that someone stumbled on it. Lawyers in the Alder Hey case called me and kicked up a stink about the illegal autopsies. In 2007, the government launched an inquiry into how the autopsies came to be carried out.[11] It eventually emerged that they had been routine at Sellafield since the 1950s. Illegal, but routine. The government duly issued an apology in 2010, and Schofield was posthumously disgraced. They even took his name off the front of a building in a science park near Sellafield that commemorated his work. But, strangely, the staggering contamination of Sellafield workers found by Schofield and reported by Popplewell has been quietly forgotten.

▌▌▌▌

In all my years of reporting about Windscale and then Sellafield, I had never been there. I decided to change that. West Cumbria is one of the most isolated and marginalized parts of Britain. It is a long way from anywhere. That was part of its attraction when scientists were looking for somewhere to make plutonium. But this isolation, coupled with the institutionalized secrecy of its largest employer, has created a climate of suspicion and a landscape of secrets. The periodic scandals and revelations have heightened the concerns without resolving them. The fact that the site had moved from military to largely civilian activities—the mandarins of the UK Atomic Energy Authority replaced by the commercially minded British Nuclear Fuels—had barely changed things.

To chart the contours of this landscape, and how it has developed over the years, I began by meeting Martin Forwood, who runs a small group called Cumbrians Opposed to a Radioactive Environment (CORE). Soft-spoken and bearded, Forwood is a former soldier, policeman, and government scientist. He took me on his "alternative tour" of Sellafield's hinterland.

We started on the banks of the River Esk near Newbiggin, a hamlet about seven miles from Sellafield. Forwood donned rubber boots, got his trusty Geiger counter out of the car, and headed for the tidal salt marshes below a bridge carrying the railway line to Sellafield. A train crossed the bridge, loaded with high-security flasks containing spent fuel from some distant power station and destined for the reprocessing works. As Forwood bent close to the mud, his Geiger counter began to click. There were radioactive particles beneath our feet. The surface of the mud registered three or four times the level in the air. Then, as he pointed his counter at mud on the exposed and eroding riverbank, the clicking accelerated to a mad chatter at around thirty times background.

The eroded mud was significantly radioactive. "Livestock graze here and fishermen dig for bait," Forwood said. "You get walkers and holiday-makers with their kids digging in the mud here." A few yards away, the Cumbria Coastal Way offered a shortcut through the mud at low tide. He pointed to green sea samphire growing in the mud. "It is a local delicacy but it soaks up the radioactivity. There are no signs to warn anyone about the mud or the samphire. There could be particles of plutonium on my boots," he added as he threw them into the back of the car.

In theory, Forwood's boots should probably have gone for burial in the concrete-lined trenches at the nearby Drigg dump, which takes Sellafield's low-level radioactive waste. So should the mud. "But you get similar readings all along the coast here," he said. "You can't remove the whole of the Cumbrian shoreline."

Forwood's readings in the eroded mud reflected a past, a few decades ago, when discharges from Sellafield's pipeline were more radioactive than today. They contaminated the mud washed into the salt marsh on the tides. He accuses official sampling programs of ignoring this buried radio-active legacy and only testing the top inch or so. The samples may be an accurate reflection of discharges today, but they may not reflect the risks to residents and visitors messing around in the mud. Since plutonium has a half-life measured in thousands of years, "these radioactive wastelands will continue to pose health risks for a long, long time," he said.

In the real world, you might be more likely to die here from eating bad shellfish than a particle of plutonium. Even so, Forwood had a point about a landscape contaminated with ghostly isotopes for the long term. He made the point most dramatically in 2005, when campaigning against Italian spent nuclear fuel coming to Sellafield for reprocessing. As a publicity stunt, he decided to bake a "Cumbrian pizza" topped with Esk salt marsh mud and garnished with Esk samphire. He took the pizza to the Italian embassy in London and handed it to the economic counselor, with a warning note about its origins. The nervous counselor took the bait. He called in inspectors from the Environment Agency, which took the radioactive pizza to the atomic labs at Harwell in Oxfordshire. There, they quarantined it for eight years before finally interring the somewhat stale mud pizza at Drigg, Sellafield's dump for radioactive waste, in 2013.

The way that radioactive discharges from Sellafield's pipeline wash back ashore has surprised scientists. Marjorie Higham, a Sellafield scientist in the early days, told the oral history project that her bosses assured her "the plutonium [would] adhere to the mud at the bottom of the sea in perpetuity. But of course it didn't. It moves around."[12]

We moved on to the foreshore of the Esk estuary, where past radioactive slicks probably washed ashore. Forwood pointed out a house overlooking the water. Years ago, he said, a couple living in the house had grown concerned after their two dogs, who had spent a lot of time playing at the edge of the water, both died of a rare nose cancer. They wondered if the mud there might be contaminated. Perhaps the radioactivity blew off the beach into their home too. The couple sent the contents of their vacuum cleaner for analysis. The report they got back found levels of radioactive plutonium and cesium thousands of times above background levels, said Forwood.

The couple brought a case against Sellafield for damages. The publicity brought them more enemies than friends. After a vibrant trade in fish sold to visitors straight from the boat on the foreshore below their house folded, locals stopped visiting the village post office the couple ran. Someone glued their front door shut. The couple had become as toxic as their environment. They sold up and left the area.

Western Cumbria is trapped by the dominance of Sellafield. Property prices suffer because of radiation fears. Few people today choose to holiday in Seascale, the small resort just along the beach from Sellafield that was my next port of call. Not with a nuclear plant looming in the background, nuclear flasks trundling past the beach on the railway, and scary local stories to ponder over breakfast. A flier near a newsstand in Seascale announced consultations on plans for a new power station close by. Most locals wanted the jobs but they would come at a price. On a cliff above the resort we stopped outside a large pink house. Two sisters, Jane and Barrie Robinson, once ran a bird sanctuary in the garden. Then tests commissioned by a charity found that many of the birds were radioactive. It turned out they often roosted in contaminated buildings at Sellafield, and may have fed on insects and algae around its open-air fuel storage ponds.[13]

Sellafield's operators had known all along about visiting birds. A public booklet on the "atomic factories" had observed as long ago as 1954 that the radioactive water "does not seem to perturb the gulls, which come to the pond from the sea a mile away."[14] It never occurred to anyone that the gulls might pose a risk to the outside world. Not until 1998, that is, when the tests of the Robinsons' birds produced a hue and cry. Soon, the authorities had strangled 1,500 radioactive birds and put their corpses into lead canisters for burial at Drigg, along with topsoil, garden plants, and even the sisters' garden gnomes.

All this is worrying for the locals. As with Three Mile Island, it is hard to pinpoint actual harm, but it undermines confidence in the area. Even so, what happened at Chernobyl in 1986 rather put West Cumbria's troubles into perspective.

Chapter 14

Chernobyl

A "Beautiful" Disaster

It had been a good night out. At just after one o'clock in the morning on Saturday, April 26, 1986, Natasha Timofeyeva, a sixteen-year-old schoolgirl, was walking home from a party in the tiny village of Chamkov, in Belarus, close to the border with Ukraine. In the black sky over the forests she saw what she told *Pravda* ten days later was "a bright flash" behind the most distant chimney of a familiar landmark across the border—the Chernobyl power plant.[1]

Timofeyeva is the only person known to have seen the moment Chernobyl's reactor No. 4 exploded, in what swiftly became the most notorious industrial accident of the twentieth century. It was the moment, many say, when the nuclear dream died; the moment, say others, when the Soviet Union died. Both statements may be true.

Inside the building that housed reactor No. 4, nobody on the night shift saw the flash. But they heard a loud thud, followed by a thunderous explosion, before the lights went out. "The doors of my office were blown out," recalled one survivor, engineer Alexander Yuvchenko. "I thought that maybe war had begun. I couldn't imagine it was something to do with the reactor." But it was. The reactor had overheated during what the handful of workers on the shift had thought was a routine powering down, prior to maintenance. The explosion blew the lid off the reactor and exposed its burning uranium core to the skies for ten days, releasing vast amounts of radioactivity into the air. The Chernobyl catastrophe had begun.[2]

Reactor operator Valery Khodemchuk's body was never recovered. His colleague Vladimir Sashenok was found unconscious, breathing bloody foam. He died of radiation burns before dawn. Three more workers died after being sent to the reactor hall to manually lower control rods and shut down the reactor, only to discover that the hall no longer existed. As Yuvchenko followed them to find out the extent of the damage, he experienced a moment of sublime stillness: "I could see a huge beam of light flooding up from the reactor. It was like a laser light, caused by the ionization of the air. It was light-bluish, and it was very beautiful. I watched it for several seconds. But if I had stayed, I would probably have died on the spot."[3]

Yuvchenko suffered radiation sickness, and by the following evening had been airlifted to a Moscow hospital. After months on blood transfusions, his body was purged of radiation. Twenty years later, he still suffered ulcers from the radiation burns and needed regular skin grafts. Radiologists estimated he had received a dose of more than four thousand millisieverts. That would be enough to quickly kill most men, but he had survived.

Within minutes of the explosion, six on-site firemen rushed from their dormitory to the reactor. They had no protective gear, nor any warning about the dangers. As the wife of one, Vasily Ignatenko, said later: "They went off just as they were, in their shirt sleeves. They had been called for a fire. That was it. They tried to beat down the flames. They kicked at the burning graphite with their feet."[4] The firemen went onto the roof to extinguish burning reactor fuel that had been blasted from the inferno. By morning, all six were in the hospital, vomiting and weak with puffed-up faces—clear signs of radiation poisoning. All died within a few days from the huge doses of radiation they received that night.

At 3 a.m., news reached Moscow of a disaster at one of the Soviet Union's most far-flung nuclear power stations. Anxious apparatchiks heard that an unauthorized midnight experiment had been under way at the plant. Operators had wanted to find out whether the reactor could be safely shut down if its cooling system lost external power. Perhaps they had been reading about the similar problem at Three Mile Island seven years before. The operators figured that the residual power produced in the turbines while the nuclear reactor was shut down would maintain Chernobyl's cooling systems sufficiently to prevent the reactor from overheating. But they did not know for sure. The problem was that the experiment designed to answer the question carried its own risk of disaster.

The operators followed the plan for their experiment. As they gradually powered the reactor down for maintenance, they turned off the cooling system. They then watched and waited to see if the reactor started to heat up. They believed that if things went wrong and temperatures started rising, they would have time to lower the control rods to halt all reactions—a task they knew took twenty seconds. But they were wrong.[5]

Powering down seemed to be going well until 1:23 a.m. and 40 seconds. At that point, the fuel in the reactor rapidly began to overheat. So the shift foreman, Leonid Toptunov, immediately ordered that the control rods be lowered. But after just four seconds, the reactor became so hot that some graphite channels that held the fuel ruptured, raising temperatures further. At 1:23 and 48 seconds—just eight seconds into the shutdown cycle—there was a violent explosion, followed by another a few seconds later.

The exact sequence of what caused the explosion is to this day unclear. The official version is that water in the cooling system boiled. Some American nuclear engineers think the botched experiment triggered some kind of a thermal shock that blocked valves in the cooling system, after which escaping water caused the reactor to go critical and catch fire.[6]

Whatever the precise chain of events, the two blasts were so violent that they dislodged a ten-foot-thick concrete lid on top of the reactor. This lid, which weighed more than two thousand tons and was all that separated the reactor core from the world outside, ended up almost vertical. With temperatures soaring and air rushing into the blazing reactor core, more blasts occurred. Chunks of burning fuel and graphite erupting from the reactor began raining down on the roofs of surrounding buildings. Dozens of fires took hold, each releasing radioactive gas, dust, and lumps of fuel into the night air.[7] Among the debris were isotopes with half-lives of only a few seconds that released great surges of radiation in the immediate vicinity of the power station. Other material, with longer half-lives, spread farther and remains dangerous in the fallout zone around Chernobyl to this day.

||||

The operators at Chernobyl that night had no idea what to do, other than fight the fires. They also little understood the perils of the smog of radiation all around them. In the hours before dawn, sixty-nine more firefighters from surrounding towns and military barracks joined their comrades from the plant fire brigade on the reactor roof. Some had dose meters and wore lead-lined suits, but many were equipped with no more than pathetic gauze face masks.

The following morning, Moscow's elite of nuclear engineers and scientists leapt into action. One advantage of the Soviet system was that everyone followed orders. So by lunchtime, a shock troop of the country's best nuclear scientists were on planes to Chernobyl. They were headed by Valery Legasov, of the Kurchatov Institute, the Soviet Union's top atomic energy laboratory. Arriving at the scene that evening, he remembered "a crimson glow that expanded to fill half the sky. We could see a white pillar several hundred meters high consisting of burning products constantly flying from the crater of the reactor," he told *Pravda* two years later.[8]

Legasov learned that water had failed to douse the fires. The uranium fuel just kept burning. So they decided to try smothering the flames by throwing material on top of the reactor core to starve the fires of oxygen. That proved a hard job. In the next ten days, some 1,800 helicopter flights dumped more than five thousand tons of sand, clay, the chemical boron, and lead onto the burning reactor core before the flames subsided.

By that time, it is reckoned that twenty-one plant operators and firefighters had received doses of more than six thousand millisieverts. Of them, all but one died of acute radiation poisoning in the weeks after the accident. They included twenty-five-year-old Leonid Toptunov, the shift foreman who had ordered the lowering of the control rods, as well as firefighters and some of the helicopter pilots, who received heavy doses as they hovered above the inferno. Another seven people who received more than three thousand millisieverts also died.[9]

What of the tens of thousands of people living close to the plant? They too received dangerous amounts of fallout in the first few hours after the accident, many of them while still in their beds. The response of health authorities was often criminally slow and chaotic. Local doctors did not know to look for symptoms of radiation sickness. There was little iodine available to hand out as a prophylactic against the heavy emissions of radioactive iodine that causes thyroid cancer.

Evacuations were patchy. At Legasov's insistence, the fifty thousand inhabitants of the model Soviet dormitory town of Pripyat, just two miles from the plant, were evacuated on Sunday, the day after the accident, when dose rates are reckoned to have been one millisievert an hour, the equivalent of almost nine thousand millisieverts in a year. They were given two hours to collect their belongings and get into more than a thousand buses commandeered for the purpose. With many allowed to leave in their cars—nobody had considered how the contaminated vehicles might spread radioactivity across the country—a convoy ten miles long formed on the road toward Kiev, the Ukrainian capital about a hundred miles to the south.

But the fourteen thousand inhabitants of the other major settlement, the town of Chernobyl ten miles from the plant, were not evacuated for another week. They were joined by the residents of more than a hundred small villages in the forests around the plant—mostly in Ukraine, but also in Belarus and a handful in Russia—as the authorities declared a twenty-mile exclusion zone around the plant. By the time these people were taken away, the doses were much lower. The damage had been done.

Over the ten days that the reactor burned, roughly a third of the radio-activity in the reactor was released into the atmosphere—an estimated 150 million curies. That was a hundred times the releases from the Hiroshima and Nagasaki bombs combined, seventy-five times more than in the cloud from the 1957 Mayak explosion, and three thousand times more than the Windscale fire. Fallout was most intense within a mile of the reactor. It killed trees in nearby forests and accumulated in the marshes along the Pripyat River. Farther away, it spread with the winds, forming a series of tongues of intensive fallout radiating from the reactor. The exclusion zone only vaguely reflected the concentration of contamination. As a result, many heavily contaminated villages in Belarus were not emptied until months later.

It seems the Pentagon knew about the accident within minutes but didn't let on. By chance, it had a satellite passing over Ukraine as the accident unfolded. Sensors revealed heat, fire, and radiation. Initially, analysts thought the carnage below meant a nuclear missile had blown up in its silo, but when they consulted a map, they realized there was a nuclear power plant down there.[10]

The first public response of the Soviet authorities was to say nothing. The outside world beyond the intelligence community first got wind some thirty hours later, after an alarm sounded as workers at a Swedish power station north of Stockholm, a thousand miles downwind from Chernobyl, clocked in for work on Sunday morning. Some of the workers had been caught in a shower of rain that contained fallout from

Chernobyl. They triggered a radiation alert at the gate. Within a few hours Swedish meteorologists had tracked the cloud back to northern Ukraine and the story was out.

By Tuesday, April 29, Chernobyl was front-page news round the world, but still the Ukrainians knew next to nothing. That day, just one paragraph appeared in the main Ukraine newspaper, on the bottom left-hand corner of page three. It said there had been an "incident" at Chernobyl that was now "under control." No information about any radiation risks was given to the population of Ukraine or Belarus until the exclusion zone was created a week after the accident.[11]

By then there was panic around Europe. For the ten days that the fires burned, radioactive isotopes such as cesium, strontium, iodine, and plutonium poured from the burning reactor and headed on the winds across much of the continent. Wherever it rained, there was fallout—in the Alps, in parts of Scandinavia, and even in upland Britain. In the highlands of Wales, Scotland, and England, radioactivity accumulated in soils and vegetation and caused a ban on the sale of local lambs that lasted until 2012. In northern Sweden, the lichen had absorbed so much radioactivity that authorities culled the reindeer, though they later admitted this was an overreaction. Many Swedes love their reindeer meat, but even binge-eating the stuff would have caused little harm.

||||

Short of nuclear war, Chernobyl was the nuclear nightmare incarnate. As Don Higson, a leading Australian nuclear engineer, put it: "Chernobyl was the worst that could happen" at a nuclear power station. "Safety and protection systems failed, and there was a full core meltdown in a reactor that had no containment."[12] Published estimates of the death toll range from the twenty-nine mentioned above who were killed in the immediate aftermath to as many as a million who have or will die as a result of cancers and other long-term consequences of the radiation. I have long been bemused by this huge discrepancy. So what do we know?

While the reactor still burned, more than two hundred firefighters and plant workers were admitted to the hospital with skin burns, nausea, and vomiting. Of those, 134 were diagnosed with acute radiation syndrome and twenty-eight died within three months, including twenty of the twenty-one thought to have been exposed to more than six thousand millisieverts. With one victim whose body was never recovered from the blast, that makes twenty-nine. Doctors say many more suffered skin burns that stayed with them, as well as a range of persistent diseases of the immune, cardiovascular, and gastrointestinal systems. Many firemen suffered permanent lung damage. Depression and sleep disturbance were also widespread.[13]

In the months after the accident, an estimated 550,000 soldiers, prisoners, and others were brought in from across the Soviet Union to do the dirty work of cleaning up the mess. Known as liquidators, they were by far the largest group of people at serious risk. They received an average dose estimated at 120 millisieverts, though around twenty thousand received more than 250 millisieverts, the UN Scientific Committee on the Effects of Atomic Radiation concluded in 2008. Of other citizens, UNSCEAR estimated that 150,000, mostly evacuees, received more than fifty millisieverts, though with maybe six thousand of them receiving more than one hundred millisieverts.[14] Anyone receiving more than one hundred millisieverts can be thought to be at some risk.

There is a harrowing museum of the disaster in the buildings of the fire department in Kiev. Its displays state that "45 percent of liquidators are today dead, and 50 percent are disabled." It is unclear where these numbers come from or how many of the deaths might be attributable to the disaster. After all, in the succeeding thirty years, many would have died anyway. The trouble is that the majority of the liquidators were not systematically followed up. "None of these men was registered by name. None was checked [for subsequent health] on a regular basis. They all went back to their homes," Leonid Ilyin, a former Russian member of the International Commission on Radiological Protection, an independent organization of scientists, told me in 2000.[15]

What evidence there is about the fate of the liquidators is not reassuring. A study in 1996 found a leukemia rate four times higher than usual among Belarusian liquidators who spent more than thirty days near the reactor.[16] Another study of Russian liquidators found leukemia rates five times expected levels.[17] The World Health Organization concluded that perhaps two thousand conscripted liquidators may eventually die of cancers and other diseases caused by the radiation. If so, that would be a silent holocaust that greatly shames the Soviet system.

Spotting the number of additional deaths among the millions of people exposed to some level of radioactive fallout from Chernobyl will be harder still. The claims of some environmentalists that as many as a million will eventually die lack any scientific basis.[18] The upper limit of plausible estimates of the final Chernobyl death toll is probably the fifteen thousand suggested to me in 2000 by Vladimir Chernousenko, of the Ukraine Academy of Sciences.[19] The International Atomic Energy Agency (IAEA) has suggested four thousand, though the analysis behind this is opaque. At the bottom of the range, UNSCEAR concluded that "there is no scientific evidence of a major public health impact attributable to radiation exposure . . . apart from a high level of thyroid cancer in children."[20] Gerry Thomas, of Imperial College London, who sat on the UNSCEAR committee that reached that conclusion, told my *Guardian* colleague George Monbiot: "Chernobyl resulted in 136 hospitalisations for acute radiation sickness, 28 of these died. There have been 5000 extra thyroid cancers, of which 1 percent may die of their disease over their lifetimes (i.e., 50—and that is probably an overestimate). End of story—no other scientifically validated further effects."[21]

The thyroid cancers will have come from the radioactive iodine released in the fire. It concentrates in the thyroid gland and can cause thyroid cancers years later, especially among children, if the victims don't take prophylactic iodine tablets at the time of their exposure. The Chernobyl accident released massive amounts of radioactive iodine—a thousand times more than the Windscale fire. Properly treated, it should not kill, but most agree that there has been an epidemic of thyroid cancer among children exposed during the accident.

A particular hot spot was Gomel, a Belarusian city of half a million people roughly eight hundred miles north of Chernobyl, where thyroid cancer rates were claimed to be some two hundred times the rates in Western Europe. Gomel suffered a rainstorm shortly after the accident, which dumped radioactive material on the city. The case has been controversial because one researcher who highlighted the epidemic, Yuri Bandazhevsky, the director of the medical institute in Gomel, was imprisoned by the authorities after going public with his concerns.[22] Some posit that much of the increase he saw may be due to uncovering dormant "subclinical" tumors, which are quite common among healthy children and which nobody would know about without a screening program.

Beyond the issue of thyroid cancer, the wide disparity in estimates of the Chernobyl death toll from radiation arises largely because of a continuing scientific dispute about whether there is a dose threshold below which risks are essentially zero. I will look at that in a later chapter.

But perhaps the greatest health impact of Chernobyl on the wider public, especially among those who were evacuated, has been psychological rather than radiological. The International Chernobyl Project of the IAEA in 1990 reported outbreaks of "anxiety, depression and various psychosomatic disorders attributable to mental distress." It found that millions of people living outside contaminated parts of Ukraine and Belarus believed they had illnesses due to radiation exposure. Such perceptions, which the IAEA blamed on distrust of authorities in the wake of the disaster, were "out of all proportion" to any real health effects and were "extremely harmful to people."[23]

This combination of distrust and mental distress came out strongly too in the oral history of the disaster, *Voices from Chernobyl*, by Svetlana Alexievich, winner of the Nobel Prize in Literature. One survivor, Nadezhda Burakova, said of her generation: "We're afraid of everything. We're afraid for our children and for our grandchildren, who don't exist yet . . . People smile less, they sing less at holidays . . . Everyone's depressed. It's a feeling of doom. Chernobyl is a metaphor, a symbol. And it's changed our everyday life, and our thinking."[24]

One of the victims of this psychological trauma seems to have been Valery Legasov, the top atomic scientist sent from Moscow to mastermind the official response to the disaster. He was widely blamed for what happened. Ukrainians angry at the Soviet authorities picked him out for having, as a display at the Kiev museum puts it, told leaders the reactor design "was so safe it could be built in Red Square." That was probably unfair. Legasov claimed to have been among those who had warned in advance that the reactor design left it vulnerable to accidents.[25] Certainly, after the accident, he was ostracized by his colleagues for being too candid about its causes. Perhaps as a result, he suffered from depression. He hanged himself in the stairwell of his apartment two years to the day after his arrival in Chernobyl to supervise the handling of the crisis. His previously written account of the disaster, in which he criticized the authorities, appeared in *Pravda* in the month after his death.

Months after Legasov's suicide, the Berlin Wall was dismantled. It was a bulwark against capitalism sustained by the claim that only socialism could deliver technology to meet the needs of the proletariat. It was swept away by an outpouring of public disillusion and anger about that ideal that many think, at least in retrospect, was partly triggered by horror at the technocratic failings revealed at Chernobyl. At any rate, Chernobyl symbolized those failings and set the stage for change. In *Voices from Chernobyl*, Alexievich reports on victims and refugees describing how the accident broke their faith in socialism. "Chernobyl is the catastrophe of the Russian mind-set," historian Aleksandr Revalskiy told her. "It wasn't just a reactor that exploded, but an entire system of values."[26]

Chapter 15

Chernobyl

Vodka and Fallout

The exclusion zone that has stretched for twenty miles around Chernobyl's stricken nuclear reactor since the 1986 accident is not quite the inaccessible dead zone often portrayed. Thousands of Ukrainians commute there every day to work on making safe and dismantling the plant and managing the zone itself. Yes, I needed an official permit to pass through the guarded gates on the road north from Kiev and a radiation scan before I could leave. But the scientists I was with had no trouble arranging my entry—and thankfully I was allowed to go home afterward.

First on my list was meeting some of the people who defied the government and returned to live in the exclusion zone in the months and years after their forced evacuation. Many live off the land in their old homes or have simply moved into abandoned buildings. After checking into Chernobyl's only hotel, I headed down the road to a high door that opened onto a small yard. It was opened by Markeyevych Federovych, one of the tribe of aging authority-defying returnees known locally as self-settlers. It was several hours before we were allowed to leave, a little unsteady on our feet.

Federovych, you see, is an effusive host and serves good vodka. He flavors it with herbs picked in the exclusion zone. Who knows how radioactive it is. He certainly didn't care, as we sat in his cozy front room, emptying his bottle and discussing his three decades of life as a radioactive outlaw. He was, he said, one of almost two thousand self-settlers who snuck back to their villages after the accident because they didn't like life as evacuees. They grew vegetables in radioactive gardens, hunted radioactive animals, gathered radioactive herbs in the radioactive forests, and sometimes drew water from radioactive wells. They were getting old now, but many were hale and hearty. It was good evidence, he insisted, for their claim that life was good in the radioactive zone.

Many of the self-settlers had been outlaws once before, he said, as members of the resistance movement fighting the Nazis in the Second World War. So when they went on the run in the early days of the exclusion zone, they knew where to hide to evade police and guards. Some, in their advancing years, spend the winter in cities but return in summer to live in their radioactive dachas. Some, like Federovych, live permanently in Chernobyl town, cheek by jowl with the workers and scientists maintaining the exclusion zone. Others live in distant parts of the zone; somewhere out there is a monastery of self-settling monks.

All self-settlers live in a shadowy world, officially tolerated in recent years, but outside the normal rules of state law. Some scavenge radioactive scrap metal and barter it for meat and potatoes from the clean world outside. I read before my trip that "the Chernobyl landscape is a space of exception." A research paper by sociologist Thom Davies, of the University of Birmingham, England, argued that "the sense of abandonment is matched by an intensification of social networks, unofficial risk understandings, and informal activities, making possible life within this nuclear landscape."[1]

Federovych laughed at such language. His life was not governed by academic abstractions. When the accident happened, he told me, he was a handicraft teacher at a school in Chernobyl. He joined the evacuation, taking his nine-year-old son on his motorbike to Kiev. Like many other evacuees, he took a summer break to the Black Sea, awaiting events. "But I was curious," he said. "I just wanted to see what was going on. So I visited. It was illegal, but I had a friend who was a captain on a small boat that went up and down the River Pripyat, past the power station. I borrowed a policeman's uniform and hopped off the boat when nobody was looking."

Somewhere he seems to have lost touch with his wife and children. Perhaps there were hidden motives behind his return, but if so he wasn't letting on: "I came back to my old house here in Chernobyl. It is a hundred years old and was built by my grandfather. It was sealed up. There was no water or light. So for a few months I lived in hiding, just with a few candles. But I felt at home. I soon realized there were quite a few of us doing this, both in Chernobyl and out in the villages." The police patrols guarding the exclusion zone knew about them, he said, but didn't know what to do.

The critical time for the self-settlers was 1989, three years after the accident. The government decided to clear a cadre of them out of a small, remote village called Ilinci. The police turned up en masse. But close to the village there was a military camp, and the commander there was friendly with the self-settlers. He intervened. There was a standoff, and the military won. "After that, the government's attitude changed and we became 'officially registered self-settlers,'" Federovych said, raising his glass to celebrate the triumph. "We got some electricity in our old homes."

In the early 1990s, there were an estimated 1,800 self-settlers. "But we are getting old. Now we are down to about two hundred, with fifty of us in Chernobyl town," Federovych said. "Some villages are empty again." He had no intention of departing "except in a box," he said, his clipped mustache twitching at the absurdity of his own mortality. Until then, he will take on anyone who tries to stop him living his life as he wants. Such as the policeman who had recently accosted him as he sat on the bank of the Pripyat River and told him to stop fishing because the water was radioactive. "I just told him that my father and grandfather fished here and I have fished here since I was a boy. He had no right to stop me. He went away."

Wasn't he afraid of the radiation in the fish, wild mushrooms, and berries that all the self-settlers ate? Not to mention the herbs in our vodka. No, he said. Chemical additives in the food eaten by outsiders were far more dangerous. "Anyway, look at me; don't I look healthy?" he asked. "There's nothing wrong with my fitness." He called his new wife from the kitchen and embraced her in a bear hug to reinforce the point. She seemed a little startled.

"Of course I know a lot of people who have died of radiation," he said. "But they were people cleaning up the contamination. The liquidators handled highly radioactive material. The rest of us have done fine. We only die of old age." Was this bravado? I don't think so. All the evidence is that the self-settlers are living longer and often healthier lives than the many evacuees who languish unhappily in distant towns—free of radiation but often consumed by angst, junk food, and fear. As Federovych leapt from his chair to bid me goodbye with another bear hug, I could not deny it. After three decades consuming the radioactive produce of a radioactive landscape, he looked remarkably well on it.

IIII

Chernobyl town, I soon learned during my visit, has a long and interesting history. This area of northern Ukraine was endlessly swapped among competing nations. Lithuania, Poland, Russia, and the Germans have all ruled at various times. For a long period, it was a haven of religious tolerance, with Greek and Russian Orthodox churches, as well as a Dominican monastery. It had a strong Jewish community. That tradition of tolerance ended with the Stalinist purges in the 1930s. The town's synagogue was turned into a Red Army recruiting center before the Nazis showed up and began the deportation and killing of its remaining Jews. In the 1960s, Chernobyl became subject to a new faith, when Moscow chose it as the site for one of the Soviet Union's biggest nuclear power stations. It grew quickly as the nuclear complex grew. Then came the 1986 accident, which shook confidence in Moscow so much that it helped bring down the Soviet Union—propelling what remained of the town into yet another new republic, Ukraine.

The name Chernobyl, which has persisted through most of this turbulent history, is taken from a Slavonic word for the wormwood plant, which lives in the marshes down by the Pripyat River. Wormwood is a synonym in the Bible for bitterness and the wrath of God.[2] The town is certainly a bitter place these days, mostly composed of abandoned and decaying buildings on buckled and cracked streets, though whether the radioactive fallout is the wrath of God is another matter.

The town bursts into life for two weeks each year, however. At the end of April, on the anniversary of the accident, thousands of former residents return to the exclusion zone. They pray at the town's two-hundred-year-old Orthodox cathedral, which, thanks to their generous donations for refurbishment, gleams in white, blue, and gold. It was empty on my visit. But on the anniversary, the priest tolls a bell in the graveyard, with one chime for each year since. After the service, many of the congregation walk past the statue of Lenin, one of the last left in a country free from its Soviet past, to a memorial park. There, down a long path, signs are pinned to crosses that name each of the 113 villages abandoned after the accident. There are post boxes where the visitors can leave messages for other returnees.

The park is full of memorials for the accident, in which white storks are a common symbol of hoped-for renewal. My favorite memorial, however, is outside town. A piece of artless granite realism put up by fire-men commemorates their many comrades who died in the dreadful days when the burning reactor spewed radiation across the landscape. Its hel-meted heroes are depicted rushing with hoses and pumps to the stricken plant. As I looked at it, it struck me that the heroic symbolism was entirely justified. Men really did die here in pursuit of the ideals of collective en-deavor and comradeship enshrined in the memorial. But those days are past. The accident laid bare how those socialist ideals had been betrayed. How technocratic hubris had created a lethal power plant, and bureau-cratic inhumanity led to so many deaths in the aftermath.

The number of former residents coming to the annual commemora-tion declines each year. Only the number of tourists visiting Chernobyl keeps rising. In Kiev, travel agents promote tours, mostly to Pripyat, which was once a state-of-the-art socialist new town occupied by fifty thousand people, mostly the families of workers at the power plant. Pri-pyat is much closer to the Chernobyl plant than the town whose name it bears. The workers there took a short stroll over a railway bridge to the plant each day. Its modernist architecture was the envy of architects and town planners in the West as much as in the East. They marveled at its stadium, swimming pool, Palace of Culture full of Soviet murals, and city plaza overlooked by a sign bearing the town's slogan: "Let the Atom be a Worker, Not a Soldier." On the week of the accident the municipal au-thorities were putting the finishing touches on a new fair, complete with yellow Ferris wheel and bumper cars. It was scheduled to open on May Day, but never did.

In the aftermath of the disaster, the authorities initially planned to revive Pripyat. "Street lights were repaired and greenhouses erected. We grew cucumbers for a while," Sergey Gaschak, scientific director of the Chernobyl Radioecology Center, recalled as we sat in the sunshine in the plaza. He had worked on the project, which also was to involve grow-ing crops for biofuels and to ferment local alcohol. But, like the other schemes, the idea of selling Chernobyl vodka didn't take off. Today, the revival plans have been forgotten. The swimming pool is daubed in graf-fiti, the sports stadium overgrown with trees, the plaza an obstacle course of broken paving stones, and the rusting fairground still awaits its first customer. On Lenin Street, trees fifty feet high grow out of the sidewalks, their branches pushing onto the balconies of the residential blocks. Only disaster tourists come, poking about in the ruins and taking selfies.

||||

Strewn across the wider landscape beyond Chernobyl and Pripyat are the random remains of human lives that were abruptly abandoned. Many villages were obliterated by bulldozers, in the hope of preventing anyone from returning. Some buildings remain, however. The wooden sheds of an old fur farm sit next to the rusting hulk of a fire engine by the power station's cooling pond. On the road to the security gates, I found a former kindergarten, its bunk beds rusting as the rain blew through broken windows. Birds nested on shelves that once carried picture books.

Back in Chernobyl town, the school where Federovych once taught handicrafts no longer has any pupils. Its classrooms are now laboratories and offices for the Radioecology Center, which is in charge of the wider ecological zone. There, Gaschak introduced me to its stern-faced director, Sergey Kireev. He told me how, in the aftermath of the disaster, the Soviet authorities made a strategic choice about the future of the exclusion zone. Rather than cleansing the landscape of radioactive material, they decided to manage the forests, soils, and wetlands to hold on to as much of the radioactivity as possible.

This, he said, served two purposes. First, it would prevent the radioactivity from spreading on the winds or down the Pripyat River to more-populated areas, like the capital, Kiev. Second, it would allow the isotopes in the fallout—mostly cesium-137 and strontium-90, which both have half-lives of around thirty years—to decay to a largely harmless state. He might have added that it was the cheapest option, especially in a country with no great pressure on land or queues of people demanding to return to their villages.

Forests, which cover more than half of the Ukrainian exclusion zone, are central to this containment strategy. They store the radioactivity in timber, in fallen leaves, and on the forest floor and in soils. No timber is allowed to leave the exclusion zone. The only exception is logs that fuel a big boiler that I passed just outside the zone. It generates heat for buildings. Filters remove any radiation in the fumes, but the ash is "hot" enough to be classified as radioactive waste. It must be returned to the exclusion zone, where Gaschak said it was used as a fertilizer—to help the forests grow faster and hold on to the radioactivity better.

The big problem with holding on to the radioactivity within the forests is the risk of fires that could disperse the radioactivity in smoke far beyond the zone. Fire prevention is not state-of-the-art here. Kireev showed me his plans for installing a network of smoke monitors but he had no money to put the plans into action. Months later, he watched helplessly as a fire engulfed an estimated fifty square miles of forest. How much radioactivity was spread out of the exclusion zone? Where did it end up? I couldn't find anyone with an answer. Far from being a safe store for Chernobyl's fallout, the forests may be turning into a radioactive tinderbox.

Aside from the forests, much of the radioactivity inside the zone has been absorbed by the marshes that sit either side of the Pripyat River close to the power plant. A lot of the water the firemen used in trying to stop the fire ended up here. Again, this could become a problem. Gennady Laptev, a hydrologist at the Ukrainian Hydrometeorological Institute, told me that if the river flooded the marshes it could end up washing the radioactivity downstream all the way to Lake Kiev, a large reservoir that provided much of the drinking water for Kiev. The lake's sediment is already "quite heavily contaminated," he said.

More fallout is locked up in the mud on the bottom of the power station's eight-mile-long cooling pond.[3] For a long time, the zone's managers kept the estimated sixty thousand curies of radioactivity in place by keeping the pond full. But that required constantly pumping water out of the Pripyat River and up twenty feet into the pond. In 2014, as a cost-cutting measure, the government stopped the pumps. Two years later, at the time of my visit, the pond was largely dried up. Some had feared this would cause a catastrophic release of radioactivity as the drying mud was whipped up by winds. So far this has not happened.[4] As I saw, the exposed mud was being colonized by reed beds that held it in place, though a drought might change that.

Scattered across the exclusion zone too are large particles of burning fuel that fell to Earth during the fire. Official records identify eight hundred burial sites where radioactive rubble was dumped. They contain an estimated half a million curies of radioactivity. There could be many more such sites, Gaschak told me. "We don't know where, because the burying was often done in a rush. We only discover them when people start doing construction."

Some scientists say that, despite its radioactive hot spots, much of the exclusion zone might be fit for people to return to live. After all, the self-settlers don't seem to have come to obvious harm here. British ecologist Jim Smith, of Portsmouth University in the UK, who has run three European Commission studies on the environmental impacts of the Chernobyl accident, thinks so. Certainly it could be the case in Chernobyl town, ten miles from the plant, where radiation levels are typically around two millisieverts a year. He was less sure about Pripyat, which is only two miles from the reactor. Doses there average between five and ten millisieverts a year. The highest Geiger counter reading during my visit to the town was around eighteen millisieverts a year, near the sports stadium.

Smith's main proviso was that people should not eat local fish; not eat local mushrooms and berries, which both concentrate radioactivity; and not consume milk from local cows, which graze on the land, or start digging up the waste dumps.

When I raised Smith's idea of resettlement with Kireev, he strongly disagreed. "Jim Smith is looking at calcium and strontium, but for plutonium, by law you can't return. It's not possible." His concern was the possibility of inhaling or ingesting plutonium. It is certainly present at far higher concentrations than in the buffer zone around Rocky Flats in the US, where, as we saw in chapter 9, even opening up the land for hiking is controversial. Smith thought it would be easy to avoid plutonium by avoiding local foods. But Kireev insisted that the plutonium risk would continue to make it unsafe for people to return. Unlike the other two isotopes, plutonium was likely to remain in the ecosystems of the exclusion zone for centuries to come. "This is forever, or at least for thousands of years," he said.

So what should become of the zone? There are several options being discussed. Soon after my visit, Chinese companies announced a deal with the Ukrainian government for a billion-dollar solar energy farm that would cover ten square miles of the zone and generate up to a gigawatt of energy, the same capacity as the reactor destroyed in the accident.[5] Kireev backed using other parts of the zone for the storage and burial ground for radioactive waste, including spent fuel from the country's three surviving nuclear power plants.

In fact, this plan was already under way. Having fallen out with Russia over Crimea, Ukraine's $200 million annual contract with Mayak for reprocessing its spent fuel in Russia is set to end.[6] So the government was building a store for the country's spent fuel close to Buryakovka, an abandoned village in the exclusion zone. The store will also take fuel debris removed from the Chernobyl plant as it is dismantled. A new railway track between the two sites is planned. The store's manager at the time of my visit turned out to be Kireev's son. Maybe that explained his enthusiasm.

Once the store is up and running, Kireev thought the next step would be to build a permanent dump for the country's spent fuel, and perhaps for that of other countries too. He showed me a map of possible sites. The geology was good, he said, and without any neighbors to relocate and compensate, the price would also be good. He reckoned a deep dump could be built for a bit over $2 billion, against as much as $80 billion if the same thing were done outside the zone.

||||

Whatever future emerges for the exclusion zone, there is still a stricken nuclear power plant to manage. In 2016, four thousand people worked in the exclusion zone commuting four days a week down a single railway track from Slavutych, a dormitory town created thirty-five miles away to house them after 1986. They were completing a giant arch to cover the reactor. It was slid into place at the end of the year. It will protect the wider environment while dismantling of the reactor's heavily polluted remains takes place, a task that could take a hundred years.

Publicly, the idea of any future disaster during dismantling is discounted. Everything is being managed safely. But my hotel in Chernobyl thought differently. A notice in my room advised me where to find a gas mask and respirator in case of a radioactive release. It also told me the location of the nearest radiation shelter. Officialdom was plainly nervous too. At the Radioecology Center, they showed me the control room that would handle an emergency if radiation levels started to soar. Young men in fatigues sat at desks that flashed up data collected every hour from sixty-six radiation monitors across the exclusion zone. The daily graphs showed a lot of natural variability, depending on how winds distribute the radiation seeping from the soil and forests. "We have lower levels now due to snowfall covering the soil," one of the operators told me. If the worst happened, computer models could calculate where the fallout would spread in the coming hours.

One thing did surprise me. Even though the power plant is within a few miles of the border with Belarus, where much of the fallout landed back in 1986, the emergency teams in Chernobyl had no instructions to tell their counterparts in Belarus if an accident happened and a radioactive cloud was once again heading north over the border. "We are not allowed to send information direct. We'd inform the government body responsible for collaboration with Belarus," said Kireev. It would be up to the officials in Kiev to pass on the bad news to Minsk. Thirty years ago, when both countries were part of the Soviet Union, the message didn't get through for many hours. I wondered if they would do any better next time.

Chapter 16

Chernobyl

Hunting in Packs

A white-tailed eagle soared in the crisp winter air. The majestic bird was hunting for fish in the giant cooling pond whose waters doused the inferno in the Chernobyl nuclear power station in 1986. The water in the pond was radioactive. The fish in the pond—including a giant male catfish that tour guides had named Gosha—were radioactive. And so, presumably, was the eagle. It didn't seem to matter. Wildlife is blooming in the exclusion zone around the site of the world's worst nuclear accident. Top predators like eagles and wolves seem to be doing best of all. Radioactive wolves? Bring them on.

I didn't see any wolves, but they are around the zone in record numbers, Marina Shkvyria, a wolf expert at the Institute of Zoology, in Kiev, told me. She has definitely logged forty to fifty wolves in the Ukrainian half of the zone, congregated around seven dens established since the accident. That is a conservative estimate based on counting animal tracks in winter snow during brief helicopter surveys. There are probably many more radioactive wolves down there waiting to be spotted. "The exclusion zone for me is a window into the past of Europe, when bears and wolves were the bosses here," Shkvyria said. It might also be a gateway to the future. "Here we learn to understand the realities of coexistence with nature and society."

Other ecologists are equally ecstatic at the rewilding of this part of northern Ukraine. Sergey Gaschak, scientific director of the Ukraine government's Chernobyl Radioecology Center, had been watching the upsurge since arriving to help with the aftermath of the disaster in 1986. "We can't prove it yet. But I think there may be around sixty to ninety lynx and maybe even more wolves, just among the adults," he said. "Wolves are everywhere. Bears have been here permanently for over a decade. We have seen their cubs." The exodus of people had resulted in "a reinstatement of nature unique in Europe," he told me. The exclusion zone had created "endless opportunities for animals and plants to prosper." If radioactivity keeps humans away, then it can only be good for wildlife.

Visitors expecting a blasted wasteland, or animals that glow in the dark, have a surprise in store in the Chernobyl exclusion zone. Strutting around the forests of a territory deemed too dangerous for humans to linger are extremely healthy-looking wild boar, gray wolves, Przewalski's horses, hares, foxes, moose, and even a brown bear or two. The zone seems in the pink of health. Certainly, there is no better place for researchers to unravel just what damage radiation does and does not do to nature.

Gaschak took me to Buryakovka, one of the 113 villages evacuated after the accident. It was a few miles west of Chernobyl town and near the site where the Ukraine government is building its store for spent power-station fuel. We drove down a long, potholed, and overgrown lane. At the end were a half-dozen wrecked houses. They were all that remained of the village. This area was beneath one of the most persistent fallout plumes during the fire, he said. Radioactivity was today still fifty times higher than in some other parts of the exclusion zone. But wildlife had no problem.

Among the village buildings he had set up two cameras, whose shutters were tripped by passing animals. He took the memory card out of one and plugged it into his laptop. Up popped a procession of Przewalski's horses—part of a herd released into the zone in the 1990s—plus a fox, a moose, and several red deer. The other camera had recorded much more: wild boar, pine martens, dogs, wolves, foxes, raccoons, badgers, and a couple of rutting male red deer. "I've seen lynx in this village too," he said. "Once I saw six within a few minutes."

Gaschak's checklist of wildlife in the exclusion zone included fifty-nine species of mammals, including beavers and otters; and 178 species of breeding birds, including nine types of woodpeckers, four eagle species, and eight owls. Few places seemed too radioactive, he said. He had even found starlings, pigeons, and swallows nesting inside the sarcophagus built around the remains of the reactor.

Across the border in Belarus, they have found many animals congregating in abandoned villages and farm buildings. Wolf densities are seven times higher than in nearby nonradioactive reserves.[1] The government of

Belarus long ago turned its section of the exclusion zone into the Polesie State Radioecological Reserve. In 2016, to coincide with the thirtieth anniversary of the accident, Ukraine announced that it was going to do the same. I'm not sure how compatible that is with the planned radioactive waste dumps. But the World Bank's Global Environment Facility has suggested that the two countries amalgamate the two into a cross-border reserve that would cover two thousand square miles.

||||

Some say we should not take this apparent radioactive wildlife renaissance at face value. Before visiting the zone, I had gone to a museum in the Ukrainian capital, Kiev, that is dedicated to the Chernobyl accident. It told a darker story. It highlighted the Red Forest, an area of Scots pine trees close to the reactor site where all the pine needles turned the color of rust. The trees died and were bulldozed and buried. The lingering fallout makes the forest "one of the most contaminated areas in the world today." Certainly, the readings on my Geiger counter soared as we drove through the Red Forest.

The museum's displays also claimed ominously that "new bacteria and viruses" had been discovered in the exclusion zone, along with "300 abnormal animals . . . Deformities went from 8 percent to 20 percent." It gave space to a large lurid picture of mutant twin pigs—one head and two bodies. They may or may not have anything to do with radiation or the accident. It didn't say. Maybe some of this dystopian take is for tourists, and to feed anger among Ukrainians about the radioactive mess left behind by the Soviet Union's nuclear apparatchiks. But a number of academic studies do show apparent ill effects of radiation on nature in the zone.

Vasyl Yoschenko is a forest ecologist who has worked in the zone for twenty-five years. He says that young Scots pine trees that have grown in the exclusion zone since the accident are often mutant. Tall, straight trees have been replaced by more bushlike trees, with many branches rather than a single tall trunk.[2] Canadian ecologist Timothy Mousseau and his Danish coresearcher, Anders Moller, have reported that more radioactive places in the zone suffer a range of less visible symptoms.[3] They have found fewer bacteria in birds' feathers, damaged DNA in mice, declines in insect populations such as bumblebees, and smaller craniums and reduced biodiversity among birds.

This is a worrying counternarrative to the stories of nature blossoming. So which is right? Or could both be? I attended a workshop of ecologists who have been doing research inside the exclusion zone. Researchers there tended to optimism and questioned some of the findings from Mousseau and Moller, whom they rather disparagingly referred to as "M&M." Their own studies found plenty of bumblebees, even in the more contaminated areas. They also questioned whether Mousseau and Moller were seeing greater DNA damage in the zone at all. The pair had no good baseline data from before the accident, and there was growing evidence of widespread DNA damage in normal populations of many species. Maybe what they measured was normal.

Gaschak said some of the findings may be real but unconnected to radiation. For instance, Moller and Mousseau reported that pied flycatchers much preferred nesting in boxes pinned to trees in areas with low radioactivity and shunned those with higher levels.[4] It looked like a clear effect of radiation. But Gaschak claimed the radioactivity was irrelevant. Most of the high-radiation areas in the study were in the Red Forest. He did not contest that the birds had largely disappeared from there. But the habitat had changed hugely. After the Scots pine trees died following the accident, the area was replanted with birch forest. Pied flycatchers are well known not to enjoy birch woodland. "Isn't that an equally likely explanation for their disappearance, rather than radiation?" he asked.

There seems to be a standoff between two factions in this debate that may owe more to politics among the scientists than to science itself. I found this polarization in many aspects of research into the impacts of radiation. Researchers either look for the good or the bad, depending on their general attitude to all things nuclear. They also hunt in packs. One group finds nature blooming; the other finds only genetic damage. This tribalism is pervasive. Most scientists have reputations to defend and friends to support. To try and break down barriers, a conference on the impacts of radiation on ecosystems held in Miami in 2015 was promoted in advance as a "consensus symposium." But there was no consensus. "There is a feeling of divergence rather than convergence in current opinion," it concluded.[5]

There is clearly truth on both sides. Scientists are not liars. It would be foolish to claim radiation has no impact. It may indeed be causing subtle genetic changes. It is also possible, as pessimists fear, that such changes could accumulate in the gene pool of some species, eventually escalating to produce big ecological impacts a few generations down the line. But while wildlife probably doesn't enjoy radiation, it often doesn't much like people either. In the exclusion zone, animals generally keep away from the areas where people go, such as around the power plant and in Chernobyl town and Pripyat. But elsewhere, in the empty forests and wetlands that cover two-thirds of the zone and harbor most of the remaining radioactivity, nature mostly thrives.

The bottom line seems to be that for many species in these "badlands," the radioactive threat is more than compensated for by the freedom from humans. Biodiversity in the Chernobyl exclusion zone is greater than elsewhere in Ukraine—greater even than in the country's protected areas. It appears to be growing. For Chernobyl's wildlife, it may sometimes be a short life; but in the absence of humans it is often a merry one.

Chapter 17

Fukushima

A Scorpion's Discovery

Until 2011, Japan had abundant nuclear power. A country without fossil fuel reserves had embraced the atom with relish. Its tightly knit technocratic elite had been able to grow the country's fleet of nuclear power stations with little public opposition. Neither the accident at Three Mile Island nor the Chernobyl disaster had disturbed progress. More than fifty reactors supplied a third of the country's electricity. The Fukushima Daiichi nuclear complex, with six reactors, was one of the fifteen largest nuclear power stations in the world and had been in operation since the 1970s.

A few critics worried about whether the power plants would be safe in the face of one the country's regular earthquakes. But systems for quick shutdowns were in place, and even the possibility of offshore quakes triggering tsunamis had been addressed. The Fukushima plant, perched on Japan's low-lying east coast some 150 miles north of Tokyo, was protected behind a seawall, which, at about thirty-five feet high, seemed high enough to hold off any wave. Or that's what they thought, until the afternoon of Friday, March 11, 2011, when there was an earthquake offshore.

At first, everything went to plan. As the earth shook, the reactors were automatically shut down. But then came the tsunami, which crashed onto the shore fifty minutes after the quake. It was forty-five feet high and washed right over the seawall and through the plant. The reactors were at first unharmed, but the wave severed the power supply from the grid and engulfed diesel backup generators. With no power to pump water through its cooling system, unit 1 quickly began to overheat. Nobody had envisaged this.

By early evening, temperatures were so high that the fuel in the reactor started to melt. With no power, little could be done. Realizing the scale of the unfolding disaster, the government ordered the evacuation of people within two miles of the plant.

By the next morning, much of the molten fuel in unit 1 had oozed to the bottom of the reactor vessel. The zirconium metal in the cladding around the fuel had corroded and generated hydrogen. The accumulation of hydrogen, coupled with steam from water boiling inside the reactor, raised the pressure, risking a rupture. All this was similar to what had happened at Three Mile Island thirty-two years earlier. As then, engineers vented the gases to reduce the pressure. The gases included radioactive cesium-137 and iodine-131. But whereas the venting worked at Three Mile Island, at Fukushima it did not. The pressure continued to rise, and that afternoon, exactly a day after the tsunami had hit and with the world watching on TV, there was a massive explosion that blew the roof off the reactor building.[1]

Belatedly, engineers began pumping seawater to try to cool the reactor. That evening, the government extended the evacuation zone to twelve miles but provided no explanation for what was going on or the extent of the radiation. People were by now fleeing from areas far outside the evacuation zone. Even health workers deserted their posts. Over the following three days, two of the other three reactors went the way of unit 1. Only unit 4 escaped because it was being refueled at the time of the tsunami.

At first, most of the releases were blown out to sea. Then the wind direction changed and radiation headed for the hills behind the plant. Fear grew right across Japan. Slightly elevated levels of radiation in Tokyo tap water prompted the government to tell parents to stop their children from drinking it. The panic was all but unprecedented in a country known for its discipline and reserve. As Wade Allison, of Oxford University, who has studied the disaster, put it: "In Japan everyone knows what to do in an earthquake. Half a million people got out of the way of the tsunami. But the public had understood that nothing should go wrong with nuclear plants. Absolute safety was assured. So when it seemed that the impossible had happened, there was panic." As the Japanese panicked, many foreigners were leaving the country.[2]

This was a compelling drama. Every day some new disaster occurred. It was clear that the plant's managers were constantly being taken by surprise and had no plan for assessing fallout levels in the surrounding areas or for passing on this information to a terrified public. A subsequent parliamentary inquiry found that a Japanese culture of complacency and subservience to authority and technical experts allowed the plant's operators to sleepwalk toward a "man-made" disaster.[3] Even after the disaster, officials seemed paralyzed. Nothing in their rulebook told them what to do. The evacuations were driven more by panic than planning. In the end, 150,000 people left the area. That was many more than necessary. In consequence, lives have been unnecessarily disrupted, local economies ruined, and a once productive landscape left deserted.

The Fukushima accident is often compared with the Chernobyl disaster. The on-site chaos and failure to tell the public what was going on bore strong similarities. The number of people evacuated was also similar. But otherwise they were very different. Chernobyl was a much worse radiological disaster. Huge amounts of the radioactive material inside the damaged reactor was blasted into the air and spread on the winds. At Fukushima, by contrast, most of the debris and most of the radioactivity was contained within the reactors.

Once the panic died down, it became clear that the total radioactive releases were only around a tenth of those at Chernobyl. Moreover, the majority of those releases went either directly into the Pacific Ocean or were blown by winds heading out to sea. The area of significantly contaminated land around Fukushima was only a fiftieth that around Chernobyl. Nobody seems to have died as a direct result of the accident either. While the bodies of tens of thousands of drowned victims of the tsunami were scattered along the coast within miles of the reactors, there were no early victims of the radiation releases or of the explosions in the reactors, and may never be. Bizarrely, on the fateful day when the ocean invaded, Fukushima turns out to have been one of the safer places on the Japanese coast.[4]

||||

What was good for the wider environment has made clearing up the mess inside the reactors much harder. Even finding out the state of their damaged cores has been dangerous and testing work. I went to talk about this with Yuichi Okamura, the man in charge of managing much of the site, at the Tokyo headquarters of TEPCO, the plant's owners. The three reactors that suffered a meltdown still contained dangerously damaged and radioactive fuel. Only robots had been inside, and several of them lasted for only a few minutes before radiation destroyed their electronics.

"Unit 2 is the most troublesome," said Okamura. Some 175 tons of molten fuel ended up on the bottom of the pressure vessel, "and some had leaked out."[5] A "scorpion" robot sent crawling into the heart of the reactor in early 2017 measured dose levels of up to 530,000 millisieverts an hour in one part of the core. Anyone foolish enough to go inside would receive enough radiation to kill them within thirty seconds.[6]

The exposure to individuals working around the reactors in the days and weeks after the accident were orders of magnitude lower than among liquidators at Chernobyl. Even so, 174 workers received their lifetime allowable dose of radiation and had to leave the site. For two of them this was because they wore the wrong boots while wading through pools of radioactive water outside the reactor buildings. One worker, who had suffered a dose estimated at only twenty millisieverts, received government compensation after he developed leukemia. Doctors say it is far from clear if that was radiation-related, but in early 2017 he sued the company for further financial redress.[7]

Cleaning up the mess inside the reactors will take decades. The Japanese government wanted spent fuel to be removed by 2020, in time for the Tokyo Olympics, Okamura told me. But with the environment inside the reactors too dangerous even for robots, that is most unlikely, he said.

A big challenge is managing the radioactive water. Engineers continue to pour water into the reactors to keep the heat down and prevent further explosions. This will continue for the foreseeable future. In a normal functioning reactor, cooling water is kept separate from the fuel. But with much of the fuel a molten mess on the reactor floor, that is not possible. The cooling water quickly becomes very radioactive.

Worse, there is other water entering the reactor. Behind the power station there are mountains, and rainfall off the mountains is constantly flowing underground into the station site. Before the accident, it was captured by drains and channeled into the ocean. But after the accident, the drains stopped draining, water levels in the ground rose, and water began to flow into the damaged reactor buildings. It still does. Engineers have been trying to halt the flow of mountain water into the reactors by freezing the soil around the reactors. The frozen wall extends for a mile, and to a depth of one hundred feet. Nothing like it has been attempted anywhere else in the world. Even so, the flow off the mountain constantly breaks the barrier. Okamura does not expect to stem it before 2020 at the earliest.[8]

Water that goes into the reactors eventually comes out, full of radioactive isotopes. In the days after the accident, much of it ended up in the ocean. Now, said Okamura, "we are pumping it out and treating it as best we can." Treatment plants around the site strip out most of the radioactive isotopes, but so far tritium has defeated them. It is accumulating in a growing bank of tanks. During my visit, those tanks were holding 540 acre-feet of tritium-contaminated water, the equivalent of roughly 270 Olympic swimming pools. Tritium has a half-life of twelve years, so one option is to just keep it there till it decays away.

Okamura sees the work to make the Fukushima plant safe going on for forty years at least. The cost keeps rising. By late 2016, the anticipated total cleanup cost had quadrupled to $180 billion. Some 40 percent of that was for compensation of evacuees; 25 percent was for decontaminating the exclusion zone; and the remaining 35 percent—or $63 billion—is for cleaning up the stricken plant itself. That is a lot of money.

Chapter 18

Fukushima

Baba's Homecoming

Japan's Highway 114 may not be the most famous road in the world. It doesn't have the cachet of Route 66 or the Silk Road or the Pan-American Highway, but it does have one claim to fame. It passes through what, since 2011, has been regarded as one of the most radioactive landscapes on the planet. For forty miles, from the Japanese city of Fukushima to the nuclear power plant known as Fukushima Daiichi, it winds around forested mountains, through abandoned villages, and past overgrown paddy fields where some of the fallout from the meltdowns at the power plant in March that year fell to Earth.

The highway was largely empty on my visit. The only travelers were cleanup workers and former residents carrying permits to make day trips to their former homes. My guide was Baba Isao, an assemblyman representing the people of the evacuated town of Namie, which lies close to the stricken power plant. His former home turned out to be close to one of the most contaminated stretches of the road. I kept track of the radiation levels on our journey using both the digital counters helpfully erected at the roadside and a counter I borrowed from a hospital in Fukushima. They showed gamma radiation, the main concern for anyone not eating contaminated food. The levels I logged by and large did not require protective clothing. But they were too high, according to the authorities, for permanent occupation of the exclusion zone along much of the road, especially in the mountains.

But radiation was just the start. What I found most insidious on my journey down Highway 114 was the psychological fallout of the accident. I was passing through a landscape impregnated with fear as much as radiation. This toxic mix is the backdrop to a drama of social breakdown and trauma that seems set to leave the area deserted long after it is radiologically safe for people to return. For while the radiation—most of it from cesium-137, an isotope with a half-life of thirty years—is decaying or being cleaned up, the half-life of the fear may be longer. And, say local doctors, the longer people are prevented by fears of radiation from going home, the worse the trauma may get.

Baba, a small, active man of seventy-two years, had arranged the paperwork I needed to travel down the highway. We met up at a roadside store in Kawamata. Once famous for its silk production, the small town had become a checkpoint on the border of the evacuation zone. Beyond the checkpoint lay the radioactive mountains, where heavy rains in the days after the accident intensified fallout and where the forests are especially good at retaining that radioactivity. Some forests were still contaminated enough to be categorized as radioactive waste. Past the mountains, as the road dipped down toward the Pacific coast and the power plant, were four evacuated ghost towns—Namie, Tomioka, Okuma, and Futaba.

After the accident, the government evacuated about fifty thousand people from areas where dose rates were above twenty millisieverts per year. But some 130,000 people in areas that turned out to have much less radiation also fled, either by the decision of local authorities or voluntarily. That is a lot of people—almost the same as the number evacuated from around Chernobyl in 1986, even though the contaminated area here is only a fiftieth as large.

The government plans to allow people to return to their homes when doses fall back below twenty millisieverts a year, either through natural decay or because of its cleanup. It is clear that the forested mountain areas are likely to be off-limits for many decades. But elsewhere the calculation is not easy. As my Geiger counter revealed, radiation levels everywhere are extraordinarily variable. They can change tenfold within a few yards. People understandably fear that they might return home to a hidden radiation hot spot.

As Baba and I drove into the exclusion zone, the landscape rapidly emptied of people. Houses sat abandoned and rice fields were overgrown by bushes. Soon there was no cell-phone signal. We stopped first at Yamakia. Nobody is allowed to live in Yamakia. Yet when we measured the radiation outside houses along the road, doses were only around two millisieverts per year, a tenth of the government safety threshold for reoccupation.

Maybe danger lurked close by, however. In the forests of Yamakia ecologists have documented widespread deformities to pine trees that they say are the result of radiation.

After Yamakia, as we climbed into the mountains, radiation levels did begin to rise. At Tsushima village, in the lee of Mount Hiyama, the digital display on an official Geiger counter near the school showed a dose of twenty-one millisieverts per year, ten times the rate at Yamakia and just above the limit for human habitation. We peered into an abandoned gas station, but there was no one home. "The owner went to Tokyo," said Baba. "I don't think he will be coming back." The new residents seemed to be wild boar, which had excavated the soil right by a vending machine on the forecourt. It was no surprise. On roadside shoulders, in overgrown gardens, and behind public buildings, their rooting was widely evident on my journey. We saw boars eating abandoned crops in fields. Radioactivity seemed to be no barrier to their activities.

On the evening of March 12, 2011, the day after the disaster at the power plant began, 1,400 people from Namie town came up the hill to Tsushima, after being ordered to evacuate. "I was among them," said Baba. "We had no information about what was going on. People were just told to come here. When we arrived, we went to the village police station and found that the police there were in full protective clothing against the radiation. When we asked them why, they said it was a precaution in case they had to go to the power plant. I don't think we were told the truth. They obviously knew something serious was going on that the population hadn't been told about. That's when our suspicion about the honesty of the authorities began."

During the days afterward, as the frightened evacuees sheltered in municipal offices at Tsushima, officials of TEPCO, the company that ran the power plant, also moved there. "But they wouldn't tell us anything about the contamination either," said Baba. That bred further suspicion, though most probably the officials knew little more about the radiation levels than the refugees from Namie. If they had known they wouldn't have gone there. It later emerged that radiation levels at this mountain evacuation center were higher than back in Namie.

In the years since, Tsushima had become an unofficial shrine to the disaster. In the window of an abandoned shop, passing evacuees had left a series of posters with bitter, ironic messages. Baba translated: "Thanks to

TEPCO, we can shed tears at our temporary housing," said one. "Thanks to TEPCO, we can play pachinko," offered another. One, in recognizable English, just said, "I shall return."

Back on the road, we stopped at a rice field overgrown with willows now twenty feet high. The owner had moved to sheltered housing with his wife, said Baba, who knew them well. "But he died soon afterwards." More from grief than anything else, he thought. We drove for some time without seeing anyone else on the road. Then a car stopped, and a small, bashful woman got out, bowing before us but keen to talk on this lonely road. Konno Hidiko had come from Suma, a town up the coast where she now lived. She was driving to Namie to clean her parents' old house and remove weeds from around an ancestral grave in the town cemetery. The weekend after my visit was a Buddhist holiday, and her family would be coming to remember their forebears.

"My parents are dead now, but I still clean the house. There are mice inside, and wild boar have been in," she said. "I can never return to it. We might come back and build a new house there one day." Baba took notes and, being an assiduous local politician, offered help if she needed it.

Anger, sarcasm, grief, and nostalgic reverence: Baba's emotions were coming thick and fast in this deserted land. I soon realized why. Near the brow of the next hill, Baba stopped his car and walked up a pathway swathed in vegetation toward a building hidden in the undergrowth. He seemed to have forgotten me for a minute. Then he turned. "This is my house," he said. "And there is my rice field and my mountain behind." The house was shuttered. I noticed laundry hanging at a window upstairs, still there more than five years after it was hung up to dry on the morning of the evacuation. Moving around his land, Baba showed me his plum trees: "The fruit is too dangerous to eat now. And the water in our well is contaminated too. Nature here is beautiful, but we can't fish or collect bamboo shoots or eat the mountain vegetables that people used to harvest from the forests. All these are things of the past."

In the undergrowth, we noticed soil disturbance. Wild boar had been visiting, though a trap left for them was empty.

Behind Baba's house was a small cowshed where he and his school-teacher wife had kept cattle. I checked the Geiger counter. In the cowshed, it read twenty-six millisieverts. But when I pointed it at the undergrowth round the back, the needle shot up to an alarming 80 millisieverts. That was four times the government's safe level for habitation. No wonder Baba had no plans to return. In a hay shed, we found a stash of old election banners. The politician in him stirred again. "I am just the son of a farmer. I wonder who has a right to destroy our home and my livelihood," he mused bitterly as I asked him to pose for photographs. "Please tell the world: No Nukes."

Outside the abandoned post office near Baba's house, an official monitor read fifty-six millisieverts. My counter agreed, but when I bent down and pointed it at a sprig of moss pushing through the tarmac, it went off the scale. "They measured five hundred millisieverts here last week," Baba said. "Moss accumulates radioactivity more than almost any other vegetation."

We drove on. Every few miles we passed massive pyramids of black plastic bags laid out in fields. They contained radioactive soil stripped from the roadside verges, rice fields, forests, and gardens of Fukushima, as part of government efforts to decontaminate the land. I wasn't clear which places had got "cleansed" and which were simply left. It seemed somewhat random, and nobody was very happy with the progress. For Baba it was too little, too late. While the government publicly boasted of "cleansing" the forests, this normally amounted to clearing trees and stripping soil for a distance of eighty feet around houses and roads. "It's not enough," he said. "When the wind blows, the radioactive soil comes into our houses. We don't trust the government on this."

Other people I spoke to thought all attempts to cleanse the forests were foolish. It was extremely expensive. In places, it unnecessarily exposed to radioactivity thousands of cleanup workers, many of them casual laborers and homeless people from distant parts of Japan. In others, it simply was not necessary, I was told by scientists at Fukushima Medical University. That amused me, because the authorities of the same university were themselves stripping soil from campus baseball and soccer fields, even though the university was far from any exclusion zone and with minimal fallout. The black bags surrounded the main university building. The embarrassed scientists told me that the authorities wanted to placate parents who might be fearful of sending their children there. They had no secret knowledge about hazards, they insisted. I believed them. But it didn't look good.

At the time of my visit, there were an estimated three million bags of radioactive soil littering Fukushima prefecture, all neatly tagged and piled on blue tarpaulins. Eventually, it all had to go somewhere. Facilities for storing, incinerating, and burying the radioactive soil were planned. But the task of transporting the soil was so great that the authorities were building a new road so the trucks could bypass the scenic mountain villages along Highway 114. No wonder the estimated eventual cost of cleaning up the land had been rising. By late 2016, it had reached $45 billion.

Driving on, we passed a second checkpoint. We were now approaching Namie, just three miles north of the power plant. I was intrigued about what I would find at Namie. Just before my visit, I had seen photographs of the town published in mainstream media such as the *Guardian* and CNN. The photographer claimed to have recently braved high radiation levels to sneak into the ghost town. In his images, he posed wearing a gas mask, to show how dangerously radioactive it was.[1]

My experience had been very different. My visit had required a request in advance but no subterfuge. And I found Namie a surprisingly busy ghost town. Nobody was yet allowed to live there, but radiation levels were low enough to allow a return. Around four thousand people worked there every day, repairing the roads, refurbishing the railway station, building new houses, and knocking down quake-damaged shops, ready for a planned return of its citizens.

The scene was eerie enough without adding silly props. There was plenty of earthquake damage and vegetation pushing through cracks in the roads and forecourts. Many shops, hairdressing salons, and other commercial premises were just as they had been left. Abandoned bicycles stood upright in shelters that still kept off the rain. Japanese order and propriety largely prevailed. The traffic lights functioned and truck drivers obeyed them; I bought lunch at a 7-Eleven and the vending machines had Coke in them. Nobody wore protective clothing. Workers had Hi-Viz jackets, not gas masks. My biggest safety concern was the news conveyed breathlessly over the town's public-address system just after lunch that a bear had been spotted in the suburbs.

Radiation levels I measured in the town were down to around two millisieverts a year. That was a tenth of the threshold set by the government for return. Lower, in fact, than I recorded in Fukushima City, which was never evacuated. Baba claimed there were still hundreds of hot spots in the town where doses might still exceed the threshold. Perhaps so, but I didn't find any.

The authorities had moved into the ground floor of the old city hall to plan the citizens' return. The question remained: would they come? Some older people would no doubt want to see out their days in their old homes if they could; and some people felt they had little choice. Around three thousand of the town's former citizens were living in cramped temporary shelters and trailer parks across the prefecture. "The conditions are terrible; you can hear your neighbor open the refrigerator," said Baba. "But many have a lot of misgivings about returning, and I think their fears are totally justified."

Baba certainly did not see it as part of his job as an assemblyman in the town to encourage their return. This was partly a reaction to his own situation, with an uninhabitable home in the badly polluted mountains. "It is totally unthinkable for me to return to my old place, so I cannot encourage them to return to theirs," he said. But he spoke what others felt.

A survey of the twenty-one thousand former residents of Namie had found that only 18 percent hoped to return.[2] Other towns where people had already been allowed back had seen a desultory response. Naraha was declared open nine months before my visit, but most former residents had stayed away. Namie's chamber of commerce had 620 members in 2011, but only twenty businesses remained registered, said Baba as we stood in front of a half-demolished jewelry store. "There is a vicious circle. No people means no business, but no business means no people will return."

On my return from Namie, back outside the exclusion zone, I visited an evacuation camp near Kawamata. The prefabricated cabins were gradually emptying as their temporary residents returned home, were resettled by the authorities, or simply bought somewhere else. I met a group of old people still waiting. They said they wanted to go back to their former homes in Namie and elsewhere. They would do so as soon as the restrictions were lifted. Their children, on the other hand, especially those with young families, feared the potential effects of radiation. "Young people simply won't go back," one told me. "Before the accident we were a big family, living close to each other. Now we are scattered. We old people are not worried for ourselves, but we are worried about our grandchildren."

In the end, perhaps the prevailing view was summed up best for me by Ito Tatsuya, a former teacher whose wife had run a pharmacy close to the station in Namie. However much money is spent trying to decontaminate and rehabilitate the town, he said, "it's a strange feeling, but I never want to go there again."

What is also true is that the longer people stay away, the harder it is to go back. In the years since the accident, most evacuees have moved on in their lives. They have new homes and new jobs. Their children are in new schools. Surely that, at least, is no bad thing. At a kindergarten in Soma, a city close to the exclusion zone where many of the evacuees ended up, they told me there was a baby boom going on in the town. The new intake of children was the biggest since the accident. Doctors say the birth rate is rising in many such places. Life, it seems, goes on. Even so, the badlands around the power plant may stay largely empty long after the radiation has gone. Highway 114 will remain deserted.

Chapter 19

Radiophobia
The Ghost at Fukushima

Shunichi Yamashita knows a lot about the health effects of radiation. He was born in Nagasaki in 1952, seven years after the world's second atomic weapon obliterated much of the city. He was brought up in the city and still lives there with his mother. "She was sixteen years old when the bomb dropped, and she was two miles away," he told me when we met at his office at the university. "She's still alive and lives in my house. She has had lots of diseases, including leukemia and tumors, but she has a very strong heart and is still going at eighty-eight years old."

Yamashita has spent his professional life researching the health of survivors of the Nagasaki bomb and other victims of nuclear atrocities, becoming a world expert on radiation and thyroid cancer in the process. He has always wanted to put his expertise to the best use. "I went to Chernobyl in 1990, just as the cancers started arising there," he said. He has returned there more than a hundred times, finally concluding that, with the exception of thyroid cancers, the health effects of the accident on ordinary citizens have been small. Then, when a tsunami disabled the Fukushima Daiichi power station in 2011, he rushed to the scene, taking a post advising the local authorities about radiation and helping to mastermind a health survey of evacuees.

But two years later, nobody wanted his advice. He lost his official advisory post because of "poor communication" and was all but drummed out of the prefecture.[1] "Yamashita attracted all the opprobrium," said Keita Akakura, the PR man at Fukushima Medical University, who has seen a wave of attacks on medical professionals since the Fukushima accident. "In the world of the Internet here, he's a demon."

Two things went wrong for him. First there was disagreement about when it would be safe for people to return to their homes. After his experiences in Chernobyl, Yamashita thought it would be best to get people home as soon as possible. So he recommended the government set a threshold for returning of one hundred millisieverts a year. He said this figure was supported by the International Commission on Radiological Protection, an independent network of radiation scientists. That was true, but the ICRP gave a range of options, saying sometimes ten millisieverts might be more appropriate.[2] The government settled on twenty millisieverts. "Afterwards, many people complained that I wanted people to stay in dangerous places," Yamashita told me.

"But probably his biggest sin in many eyes," according to Akakura, "was to tell people to smile—that if they smiled they were less likely to get sick. He thought that message would be understandable to the public. But it seemed like he was trivializing radiation, and it increased the suspicion that he was in league with the nuclear people."

Yamashita couldn't resist a rueful grin when I asked him about the smile episode. "It was at a public meeting, ten days after the accident. Everybody was very stressed. I said don't worry too much. Relax. Smile. My audience understood me. They realized the importance of relaxation. It would boost your immune system. Nobody reacted at the time, but later it was cut and pasted: people used it to attack me." He spoke sense but became a lightning rod for discontent about the accident. "The population lost trust in the experts," Yamashita said with a sigh.

Fukushima has become a case study in how a relatively small amount of radiation can still create a climate of fear, and why the collapse of trust in experts is such a danger in the nuclear industry. Because you can't smell or see or touch radiation, trust in experts is essential for people to make sensible decisions after an accident. But if that trust is lost—which is likely after a nuclear accident—then it doesn't make much difference how much radiation is lurking in the air, or water, or soils, or how truthful the experts are being. Because people have no way of verifying the truth of what they are being told.

There was widespread panic in the days after the Fukushima accident. A big part of the reason was that the government gave out no information about radiation levels. The experts at the power station and in government were silent. People had no idea if they were going to die. Even medical doctors who knew about radiation were in the dark. "We had no information for ten days. It was a big problem," said Koichi Hasegawa, a doctor at Fukushima Medical University.

In the information vacuum, fear and rumor stalked the land. Professionals could not be professional. Fearing for their own lives, nursing staff and care workers deserted their patients. Doctors from outside feared coming to help. Even the Red Cross withdrew its staff, abandoning hospitals and evacuation centers where thousands of victims of the tsunami and evacuees from the nuclear accident were gathering. "We had had no training in how to operate after a nuclear accident, and it was frightening to deal with that kind of invisible threat," the Red Cross's Fukushima chief of operations, Shoichi Kishinami, explained later. He feared for his staff.[3] Food and other rescue supplies sent from Tokyo for the survivors of the tsunami never got delivered because nobody would drive them into the radioactive "danger zone."

Scientists tried to fill the information void. The first maps of radiation in the prefecture were produced not by the government or TEPCO, but by FMU and its neighbor, Fukushima University. "We had to borrow instruments," remembered Kenji Nanba, director of Fukushima University's new Institute of Environmental Radioactivity. "But by the end of March we had mapped the whole prefecture outside the exclusion zone, where we were not allowed to go. Some places, like Iitate village, were only evacuated after they saw our data." But while alerting some to danger, the maps showed for the first time that many parts of the prefecture were largely safe.

By then it was too late, however. Many people, traumatized by the events during the days when government officials and nuclear engineers abandoned them, will never listen to nuclear experts again. And that includes radiation doctors. Some call this "radiophobia." But that is to blame the victims when the real problems lie elsewhere. As Yamashita—an expert caught up in the tidal wave of mistrust—told me, public fears about radiation and the lingering effects of the Fukushima accident are real and corrosive. They cannot be dismissed as a phobia.

||||

This tragedy continues to play out. The biggest concern I heard from people across the Fukushima prefecture—and more widely in Japan— was about the health effects of the radioactive iodine-131 released by the plant's operators as they grappled to forestall a meltdown. That concern is not unreasonable. Iodine-131 has a half-life of only eight days, so it doesn't stick around. But if ingested—for instance, in milk from cows grazing on contaminated pastures—it concentrates in thyroid glands and can cause thyroid cancer that emerges within a few years. Children are at special risk. The only prophylactic is to give exposed people tablets of nonradioactive iodine to flood the thyroid gland and prevent uptake of the radioactive version.

Iodine-131 was in the cloud created by the Windscale fire. There was an epidemic of cases of thyroid cancer after the Chernobyl accident in Ukraine, where iodine emissions were high and few efforts were made to protect people. It was also released during the Fukushima accident, though only about a tenth as much as at Chernobyl. Because most of the fallout initially headed out to sea, the doses to humans were even less than that.[4] Japan also acted promptly to prevent iodine uptake. Besides rapid evacuation, contaminated milk and other foodstuffs that could harbor iodine-131 were quickly withdrawn from sale. So any ingestion at the time should have been well below what might cause disease. Iodine tablets were also issued.

So doctors have been fairly confident that there should not be an upsurge of new cases of thyroid cancer. Nonetheless, health authorities commissioned FMU to conduct a long-term screening program among Fukushima's children. To start things off, the doctors organized an initial ultrasound screening of thyroid glands two years after the accident—well before any cases of thyroid cancer from Fukushima radiation might show up. The idea was to use the data from this first mass screening as a baseline. It would allow them to check later to see if there had been an increase.

The trouble was that the initial screening found large numbers of cysts and nodules on thyroid glands and 113 cases of thyroid cancer. That compared with the four cases that doctors would normally expect to diagnose in such a population. Not surprisingly, there was an outcry, with banner headlines proclaiming a thirty-fold increase in thyroid cancer rates. Almost everybody I spoke to during my visit had heard of the finding, and most laypeople concluded the worst. That included Baba Isao, the assemblyman who took me to Namie. "If you say radiation doesn't cause a problem, how can we have so many cases?" he asked me. "The government can't give an answer to that question." Well, actually it can and does.

What was going on? I asked the doctors in charge of the screening. The problem, they said, is that what you look for, you tend to find. Thyroid abnormalities, including some cancers, are much more common in the general population than the number of cases actually diagnosed as a result of people going to their doctor with a concern. "We know this because autopsies reveal many nascent cancers," said Ken Nollet, the American director of radiation health at FMU. So, if you run a screening program, you are almost bound to uncover lots of thyroid disease, including cancers, that would not otherwise be known about.[5]

That is what the post-accident screening program found, said Nollet. "It's not abnormal to have cysts and nodules on thyroids during child-hood, especially during adolescence, the age at which most cases showed up." He pointed out that a Korean mass screening study using the same diagnostic techniques at around the same time on a nonexposed popula-tion found similar rates to those in Fukushima.[6]

A few researchers have contested this conclusion, notably Toshihide Tsuda, of Okayama University, who insists that the spike is "unlikely to be explained by a screening surge." He blames radiation.[7] His views have been widely quoted in the media but are largely dismissed by medical re-searchers I spoke to in Japan and elsewhere. I believe the FMU doctors. The evidence from Chernobyl and elsewhere is that it takes four years for the first cases of thyroid cancer to show up after a radiation incident. Why should Fukushima be different? The argument that the screening process showed what wouldn't otherwise have been visible seems sound.

The doctors I spoke to at the university were distressed about their failure to communicate what they had seen as straightforward science. They were, to be honest, pretty naïve. They should have seen this one coming. Nonetheless, in the climate of fear and distrust created by the in-formation vacuum in the immediate aftermath of the accident, they were unlikely to be believed, whatever they said. As Nollet said, "It's very dif-ficult to convince the public about this. And when we try, we are seen as complicit in nuclear power. They don't accept the scientists' view because they see us as nuclear allies."

If most medical researchers believe that there will be no surge of thy-roid cancer cases, what about leukemia and other cancers, which take much longer to show up? Here, there is more contention among the ex-perts. The radiation received by the public as a result of the Fukushima accident was trivial. "Less than 1 percent of the affected people got a dose of more than one millisievert from external exposure," said Sae Ochi, a pediatrician who came to Fukushima as a volunteer after the accident. "Even eating a wild boar stew wouldn't give you more radiation than a one-hour aircraft flight." Nollet agreed, saying, "Almost nobody received more than three millisieverts," which is roughly the global average annual background dose from natural sources of radiation.

Some believe any dose adds to cancer risks. So they say a few hundred people scattered across Fukushima could eventually die from the extra radiation. That is pretty theoretical. There is no scientific evidence of any increases in cancer among populations exposed at such low levels. When I met British medical radiation expert Gerry Thomas at her office in Imperial College London, she told me flatly: "We don't expect any health effects from radiation at Fukushima."

I look at this debate further in the next chapter, but here is the rub, and here is where I think nuclear authorities and their apologists should not be let off the hook: There have already been real and profound health consequences from the Fukushima disaster; but they have not been the result of the direct effects of radiation. They are due to the evacuation and stress, the unemployment and general social breakdown that the accident caused.

The evacuation was chaotic, so chaotic that somewhere between forty and sixty old people died either through being left behind in nursing homes and unattended in their own houses or because of the trauma of their removal. "One woman called the ambulance from within the exclusion zone because she needed regular oxygen therapy and nobody had come to give it. Others died at home of dehydration and hunger," Ochi told me.

A second wave of deaths resulted from depression among evacuees, she said. In late 2016, there had been approximately eighty-five suicides linked to the distress caused by the accident and evacuation. Many more people have fallen into nonlethal depressions. "It is post-traumatic stress disorder," said Masaharu Maeda, the head of the department of disaster psychiatry at FMU, whose researchers had conducted interviews with half of the evacuated population as part of a mental health survey.[8]

One study, by Yasuto Kunii of FMU, found that "the nuclear accident seriously influenced the mental health of the residents." He recommended "prompt and appropriate support, [including] continued psychosocial in-

tervention for evacuees."[9] These are real psychological and social consequences of the accident—ones that, in my judgment, are just as much
the responsibility of the operators of the power plant as any radiological
consequences.

During my visit, people expressed this psychological dislocation to me,
in many forms, from Tatsuya's "strange feeling" that he can never return,
to the grandmothers' fears for their grandchildren going near to the plant.
Ryoko Ando, who was evacuated from the coastal town of Iwaki, wrote a
paper for a medical journal about the dislocating effect of "unfamiliar values invading our lives . . . we found ourselves drowning in numbers [and]
strange units we have never heard of, such as the Sievert. Radiation had no
role in our consciousness until then; suddenly we found that it was part of
our lives, without having a yardstick to gauge it and form a judgment."[10]

The journal *Nature*'s estimable chronicler of the aftermath of the accident, Geoff Brumfiel, reported on the life of Kenichi Togawa, a keen judo
fighter who lost contact with his martial arts group after being evacuated
from Namie. He now "exercises less and rarely socializes. He drinks more
and has put on weight," spending long hours in front of a video console in
his tiny evacuation apartment, drinking *shochu*, a strong Japanese liquor.[11]

Parents fear their children will get thyroid cancer. A quarter of young
girls surveyed felt they might not be able to have a baby because of the
accident. "People with very negative views about the risks of radiation
are more likely to be depressed and to suffer feelings of self-stigma. It's a
vicious circle," says Maeda. "And not knowing when you might be able
to go back to your home makes things worse." He called all this "the
Godzilla effect," after the 1950s film about a mutant monster created by
atomic tests.

One study looked at the fate of more than 1,200 old people who had
moved to care homes in the region around the power plant in the five
years before the accident. Some had been evacuated during the accident,
and some not. Those who had been evacuated were three times more
likely to have died by mid-2013 than those who had not been evacuated.[12]
It suggested, said Claire Leppold, a doctor in Minamisoma, a town near
the evacuation zone, that "the evacuation may have been more dangerous
than the disaster itself for this population."[13] Some reports have suggested
that as many as six hundred people may have died prematurely as a result
of the stresses caused by the evacuation. But the figures are confused both

by what counts as a premature death among the already elderly and by the many deaths resulting from evacuations from areas damaged by the tsunami itself.

"Of course, we had to evacuate people," said Hasegawa. "But early on, I said that the government should plan for a quick return." Apart from a few high-dose areas in the mountains, the psychological risks of staying away have long exceeded the radiological risks of returning. "People could have been returning after a month, when the iodine had disappeared," agreed Yamashita. As time passes, the man whose advice was spurned because it did not fit the public mood looks likely to be vindicated.

Chapter 20

Millisieverts

A Dose of Reason

It is hard to imagine now, but a century ago many people thought of radiation as an elixir of life. It began with the discovery by Marie Curie in 1896 that a metal she uncovered called radium gave off radiation—and that the radiation might be used to zap cancerous cells. Doctors gave the first treatments in 1906 by smearing radium salts directly onto the surfaces of tumors. The idea quickly spread that radium had wider healing powers. People demanded it from their doctors. In the 1920s, the Radium Chemical Company of Pittsburgh offered bottles of "radium solutions for drinking" to help with "subacute and chronic joint and muscular conditions, high blood pressure, nephritis, the simple and pernicious anemias." The American Medical Association offered its support. There were radium baths and facial creams. Radium turned up in chocolate, toothpaste, suppositories, as a treatment for impotence, and even as a glowing nightlight for restless children.[1]

The tide started to turn against the wonder element when a wealthy American socialite, Eben Byers, died in 1932 of cancer brought on by binge-drinking more than a thousand bottles of a patented radium concoction.[2] His body was so radioactive that it was buried in a lead-lined coffin. Then a couple of years later, Curie herself died of anemia, which doctors concluded was brought on by her habit of carrying radium around in her pocket. Horror stories soon emerged of bone cancers among women who were painting luminous radium onto clock faces.

Yet, despite these cautionary tales, radioactivity for a long time retained its attraction as a medical restorative. And one form of radiation, X-rays, was regarded almost as a toy. When I was a child in the late 1950s, shoe shops had X-ray machines next to the displays, where customers could check fittings. I remember that a sign advised against overuse, but there were no actual restrictions. I was fascinated by the images of the bones in my feet and demanded to see more. Even today, some health spas market the curative powers of the radioactivity in their waters—the radioactivity that heats the water to such an agreeable temperature. The online come-on for one spa in Jamaica promises that its waters are "some of the most radioactive in the world."[3]

So how dangerous is radiation? How much will kill you? Is there a safe level? Should the evacuees from Fukushima have returned home within weeks of the accident, or might it still be unsafe to do so? Is it time to repopulate the Chernobyl exclusion zone? These are not easy questions. For one thing, different types of radiation zap us differently. They include X-rays and ultraviolet radiation, but here we are mostly concerned with three types: alpha, beta, and gamma. Gamma rays penetrate furthest, including through skin. They pose a threat to anyone near the source. Beta radiation penetrates much less. Alpha radiation is stopped easily, even by human skin, but if you breathe it in or ingest it in food, it can stick around, potentially for the rest of your life. The standard measure of radiation dose is weighted to take account of the different risks from different types, giving greater importance to alpha radiation. This measure is the sievert. Since a sievert is rather a lot of radiation, in practice we are usually talking about millisieverts, which are a thousandth as much. So how many millisieverts will kill you?

Researchers can't go around doing experiments on people to find this out, but there is a growing amount of evidence from accidents. Take the remarkable story of what happened in the central Brazilian town of Goiania in 1987. A scrap-metal merchant was rifling through the remains of an abandoned radiotherapy clinic when he found a machine that still contained a source of the isotope cesium-137. Not knowing what it was, he took it home and split it open. He and his wife admired the pretty blue glow in their living room. Their children played with the strange machine and the neighbors came around to join in. After two weeks some of these people started getting sick. Four of the family, who had been closest to the pretty blue glow the longest, died. Doctors later estimated they had received doses in excess of four thousand millisieverts.[4] Another 250 friends and neighbors suffered significant radioactive exposure. Of them, twenty-eight required operations to relieve radiation burns, but none died of their injuries.

Something similar emerges from Chernobyl. All but one of the twenty-one firemen and liquidators who were exposed to more than six thousand millisieverts died soon after from acute radiation poisoning. But of the 216 exposed to less than six thousand millisieverts, only eight soon died. This says little about long-term consequences, but from such accidents—along with occasional inadvertent exposure in labs and factories from Los Alamos to Mayak—researchers have concluded that more than four thousand millisieverts in a short space of time will usually kill you within a few weeks, while about one thousand millisieverts will produce immediate symptoms of acute radiation sickness. That means serious damage to cell membranes, resulting in vomiting, diarrhea, and internal bleeding, and usually also radiation burns and damage to the body's immune system, opening sufferers up to all manner of diseases. Not surprisingly, deaths are also frequent in this group.

Doses a bit below one thousand millisieverts may give some of these symptoms, sometimes called chronic radiation poisoning. But below one thousand millisieverts the effects seem to be mostly long-term, with increased risk of leukemia and some cancers. That is the experience of the survivors of Hiroshima and Nagasaki. It may be the fate of many of the Chernobyl liquidators and some of the seven thousand or so people evacuated from near Chernobyl who received doses of above one hundred millisieverts before they got on the buses.

But how great are the risks at these lower levels? Can we draw a simple line on a graph, in which half the dose brings half the risk? Or might there be some threshold level beneath which there are no effects? This is still hotly debated. The safety-first assumption has long been that there is no threshold. This idea is increasingly being challenged, however, because once you get down to exposures of one hundred millisieverts or below, there is no solid evidence of any risk at all. The situation is further confused because these kinds of levels are close to those experienced by many people in everyday life from natural radiation.

||||

The greater part of the radiation that most of us receive comes from natural sources. The average natural dose is 2.4 millisieverts a year, but there is a lot of variation. In Britain, for instance, thousands of people receive annual doses above two hundred millisieverts, and some places far exceed even that—for instance, Kerala, in India.[5] These high natural doses usually come from rocks such as granite that emit radioactive radon gas, which makes up about half of the radiation exposure in our environment. Unlucky Brits get their high doses in places such as Aberdeen, the "granite city."

Another natural source is cosmic radiation streaming in from outer space. Living on the International Space Station exposes astronauts to 170 millisieverts in a year. Even aircraft fly high enough to up the doses of passengers. A flight from New York to London, for instance, delivers around 0.1 millisieverts. There has been talk in the US of introducing a legal exposure limit for aircrews.[6]

For most of us, unnatural radiation comes largely from medical procedures. A chest X-ray will probably only give you around 0.05 millisieverts; a mammogram might deliver three millisieverts, roughly a year's typical natural dose; and a CT scan can give you up to twenty-five millisieverts. Doctors target much higher doses at cancerous cells to kill them but carefully shield the rest of the body to prevent damage. By comparison, we get much smaller amounts from the lingering fallout from nuclear weapons tests half a century ago, and less again from the spread of radiation from nuclear facilities or accidents such as Chernobyl.

There are few legally enforced limits on natural radiation, but most countries set and enforce strict standards for exposure to man-made radiation. But the limits vary a lot. Typical nuclear workers are limited to a maximum of fifty millisieverts in a year. For the general public, the safety-first limit recommended by independent scientists at the International Commission on Radiological Protection (ICRP) is one millisievert a year. That is an addition of about a third to the average natural dose. But, as we saw in the last chapter, limits depend on circumstances. Sometimes, says the ICRP, it may be best to allow people to live in areas with up to one hundred millisieverts a year.[7]

These recommendations are based on estimates of cancer risk. Current estimates are that one millisievert a year may increase your chance of contracting a fatal cancer by one in twenty thousand per year, so by one in 250 over a typical lifetime. That is a small addition to the existing risk we all have of dying from cancer, which is around one in four.[8] It doesn't tell you if you will get cancer or whether the cancer you have was caused by radiation, but at least it puts things in perspective.

However, the estimate itself is contentious. It is mostly based on extrapolating from what happened to the *hibakusha*, the Japanese bomb survivors. Most received more than two hundred millisieverts from the bombs, and the group suffered a measurable increase in their cancer rate. The working assumption is that the risk from one millisievert is two hundred times less. Researchers call this working assumption the linear no-threshold (LNT) hypothesis. Note that word, though: it is just a hypothesis.

Many medical researchers say this is unduly pessimistic. They say that millions of years of exposure to natural radiation has given our bodies repair mechanisms that fight back against radiation and may protect us almost entirely from these low doses. But it is fiendishly difficult to be sure. The cancers that could result from low levels of radiation are also caused in larger numbers by other factors. How would you spot if there were a small-percentage increase in lung cancer given all the deaths from smoking? The debate is very polarized. Most researchers, I discovered, take sides in this argument: for or against the LNT hypothesis.

What is clear is that this uncertainty has a huge impact on assessing how many people might be killed by events like Chernobyl and Fukushima, or even by natural radiation. The LNT hypothesis suggests there will be big death tolls in all three cases. This is because while the extra risk to an individual from a few extras millisieverts may be tiny, many millions of people are exposed. For instance, using this approach, the US National Cancer Institute estimated that eleven thousand Americans have died or will die of cancers as a result of nuclear tests.[9]

But if the LNT hypothesis is wrong and there is a "safe" threshold below which the body can normally repair any damage from radiation, then such scary estimates may be nonsense. Deaths from bomb-testing fallout might be limited just to people with the highest doses, such as the *Lucky Dragon* fishers from Japan or the people caught in some of the nastier fallout clouds downwind of Semipalatinsk.

Figuring out deaths from Chernobyl hits the same problem. The total dose of radiation release from the burning reactor was probably between 65 million and 150 million millisieverts. Most of this was distributed among tens of millions of people, most of whom got only a tiny dose. If the LNT hypothesis is correct, then thousands of people among those millions should eventually die from the radiation. But if people who received less than, say, 100 millisieverts were not harmed, then the true figure could be far lower, and largely restricted to the liquidators.

||||

The LNT hypothesis has held sway among radiological authorities ever since a report from the UN Scientific Committee on the Effects of Atomic Radiation (UNSCEAR) in 1958 concluded that "even the smallest amounts of radiation are liable to cause deleterious, genetic and perhaps somatic effects." Yet, even then, the report conceded that "linearity has been assumed primarily for purposes of simplicity. . . . There may or may not be a threshold dose."[10]

In the years since, there has been no convincing evidence that doses below about one hundred millisieverts a year produce any measurable increase in risk of cancers, says the ICRP.[11] The body in charge of studying the Japanese bomb survivors, the Radiation Effects Research Foundation, says clear evidence of increased cancers among them starts to appear only above 150 millisieverts.[12] *Hibakusha* who received less than that seem to show no increased risk of cancers.

While, for now at least, international authorities stick by LNT, skeptics have upped their attacks. Bill Sacks, formerly a radiologist at the US Food and Drug Administration, is one. In a paper titled "Epidemiology Without Biology: False Paradigms, Unfounded Assumptions, and Specious Statistics," published in 2016, he called the hypothesis "analogous to observing that if a person takes 100 aspirins at one time there will be a single death, and then asserting that the same single death will occur on average as a result of 100 persons each taking one aspirin."[13]

Everybody accepts that there is a safe dose for medicines but that too much kills; so why not the same for radiation? It all depends, as Sacks said, on the "biological response of the organism" to radiation. He argues that evolution should have set us up well for coping with radiation, especially as radioactivity levels on the planet were once much higher than today. As another skeptic, Oxford nuclear physicist Wade Allison, put it: "If evolution had not found ways to protect life against radiation, we would not be here."[14]

Sacks, Allison, and others say this is not just theory. There is also good medical evidence to show both that our bodies will repair damaged DNA and that our immune system will kill damaged cells and replace them with healthy ones. Much of this evidence comes from seeing how our bodies deal with the heavy radiation doses that doctors inflict during radiotherapy.

Healthy people can cope with one hundred millisieverts a year with no problem, Sacks says. Other radiation scientists disagree, however. They stress not the body's resilience but its vulnerability. A single cell damaged by radiation may become malignant, they say. If it does, and if that cell isn't repaired or killed, it may spread that malignancy through the body. So there can be no "safe" level.

After some initial reluctance, I have come to believe that the threshold argument makes scientific sense. We need a rethink. While it was logical initially for regulators to play safe and presume that even the smallest doses of radiation could cause harm, the evidence accumulated since does not justify that caution. Continuing to deny the case for a threshold is in danger of becoming as irrational as the proposition that man-made climate change has not been "proved" or that we can't demonstrate that cigarettes cause cancer.

Decades of research and investigations of the victims of nuclear bombs, accidents, and workplace exposure has found no compelling evidence of an effect for doses below one hundred millisieverts. So we have to conclude that either no such risk exists, or that it is trivial compared with many other day-to-day hazards. This should not be a reason for relaxing engineering standards at nuclear plants. But it might avoid the crazy situation after Fukushima, in which most of the health effects of the accident have resulted from the trauma of a panicky evacuation and a continued life in exile.

||||

The case for applying this threshold standard to gamma radiation, which zaps us from outside the body, seems to me strong. But does it also apply to alpha radiation that gets into the body and may lodge there, attached to particles, resulting in intense radiation to a handful of cells, triggering cancers? Some believe that alpha-emitting plutonium could be a big risk here. This is the "hot particle" theory, first proposed in 1974 by Arthur Tamplin of the Natural Resources Defense Council, in Washington, DC.[15] Back then, Britain's Royal Commission on Environmental Pollution concluded that, if the theory were true, the world would have to impose limits on plutonium that would "effectively preclude the use of nuclear power."[16]

The commission decided the evidence for the hot-particle theory was slight. Government regulators and international research agencies such as the ICRP have ever since taken that view. They argue that particles in the lung don't stay in one place, but move around, limiting the chance of them giving intense doses of radiation to individual cells. They add that lab animals that breathe in these supposed "hot particles" do not show higher cancer rates. What data there is on humans seem to point the same way. In the early days of the Manhattan Project, around twenty-five workers at Los Alamos breathed in plutonium in quantities that, according to the hot-particle theory, should almost certainly have killed them. But they lived. The same was true of people shamefully subjected to experiments in which they were injected with plutonium. Their survival rates too were quite good.[17]

Back in the 1970s, one advocate of the hot-particle theory, the American activist Ralph Nader, claimed that a pound of plutonium dust released into the atmosphere and spread round the world could kill every human. Environmentalists still quote it. But it is silly. Around four tons of plutonium has been distributed round the globe in the fallout from atmospheric weapons tests, with more from processing plants such as Rocky Flats and Mayak. Geologists say plutonium has turned up in tiny quantities almost everywhere, and we are not all dead. The late Bernard Cohen, a Pittsburgh physicist, talked about the "myth of plutonium toxicity" and once offered to eat as much plutonium as Nader would eat caffeine. Nader did not take up the offer.[18]

I remain jumpy about plutonium, however. Our bodies can probably shirk off low doses of radiation with little harm. We are evolved to deal with that. But the idea of a single particle of some radioactive substance—especially one that did not exist in nature until we began irradiating uranium in nuclear reactors—being lodged somewhere in my lungs, zapping a few cells with intense radiation year after year, does sound dangerous. Is that my radiophobia? Maybe.

Part Four

Cleaning Up

THE ATOMIC AGE has created a fast-growing legacy of abandoned reactors and radioactive waste. That legacy is rapidly turning into a nightmare—a trillion-dollar liability comprising radioactive concrete and steel, fuel rods, sludges, and hot liquids. Some of it will be dangerous for tens of thousands of years. Much of it is languishing in dilapidated stores, leaking tanks, or flooded salt mines from which it may have to be retrieved. Most of it has nowhere safe to go. Germany is finding that going nuclear-free does not free it from the waste burden. On British coasts, the carcasses of nuclear reactors are likely to sit abandoned into the twenty-second century. And my home country also houses perhaps the most alarming legacy of all: more than 130 tons of terrorist-ready plutonium in a poorly protected store on the Cumbrian coast.

Chapter 21

Sizewell

The Nuclear Laundryman

On the last day of December 2006, the Sizewell A plant on the coast of Suffolk in eastern England stopped generating electricity after forty years on the grid. Cameras in the control room broadcast the switch-off to the local sports club, where there were rounds of applause and toasts from men who had worked there since the early days.[1] The plant seemed to have got to the end in good order. Its original design life of twenty years had been doubled. Now it was set for decommissioning. Its last consignment of fuel would be removed; then the plant would be put into what its engineers call "care and maintenance" before eventual dismantling to leave a greenfield site.

What could possibly go wrong? In the following days, the two reactors were depressurized and a lot of equipment was taken off-line. The final batch of five thousand spent-fuel rods was put into skips and lowered onto the bottom of a giant pond next to the reactors, while the rods slowly cooled. Then, something went wrong. At 10:45 on the morning of Sunday, January 7, exactly a week after shutdown, a plastic pipe circulating water to the cooling pond split. Water burst through a sixteen-foot gash. With water continuing to leave but no more arriving, the pond rapidly began to empty. Nobody noticed.[2]

Within forty-five minutes, the water level had dropped by more than a foot. The tops of the hot fuel rods were on the verge of being exposed to the air. If that had happened, they would have swiftly overheated and set in chain a major nuclear disaster, with radioactive cesium-137 and iodine-131 released into the air over Suffolk. Luckily, at that moment an operator went into a laundry room near the pool to sort some clothes. He noticed the floor was flooded and alerted the control room. They stopped the circulation pumps, halting the leak.

It was a lucky escape. Had it not been for someone deciding to do his Sunday-morning laundry minutes after the pipe ruptured, the leak might not have been spotted until the next routine 12-hourly visual inspection. The pond could drain in ten hours. "In this worst-case scenario, if the exposed irradiated fuel caught fire, it would result in an airborne off-site release," government inspectors wrote later. "There was a significant risk that operators and even members of the public could have been harmed."[3]

This chain of events should not have been possible. The inspectors found that the plastic pipe did not meet design specifications and had split previously, just eight months before. Worse, a recently installed system designed to spot falling water levels in the pond had not worked. Even if it had, nobody in the control room would have noticed because the alarm it would have triggered had already been sounding for two days amid what seems to have been a chaotic shutdown of equipment across the plant.[4]

The picture that emerged from the inspectors' report looked like a serious failure of basic safety systems: not a random mishap but something more systematic. You might think that the public, especially Sizewell's neighbors, should have been informed and someone held to account. Maybe there were lessons for other reactors of the same design still operating. But no. There was no prosecution; no disciplinary hearings; and the report of the government's Nuclear Installations Inspectorate (NII) was never published. Its findings only came to light two years later after a freedom-of-information request from former Sellafield scientist turned nuclear consultant John Large. He was working for local antinuclear activists who had got wind of the accident. He published extracts.[5]

Caught in a flurry of headlines, the NII's higher-ups abruptly backtracked on their own inspectors' alarming conclusion. Chief inspector Mike Weightman told reporters that his staff were wrong to say the pool could have entirely emptied. The fuel would only have been partially exposed and would not have reached the 1,100 degrees Fahrenheit at which it could have caught fire.[6] Large contests this. He says the exposed cladding around the fuel rods would have caught fire at much lower temperatures, and that would have ignited the fuel. Who is right? Do the NII inspectors who actually visited the scene of the accident agree with Large or with their boss? We don't know.

Large sent me the full version of the report. It described itself as a "preliminary report." However, even though its main findings had been brusquely set aside by Weightman, no subsequent "final" report has ever been published to set matters straight. In 2016, I asked the press officer of the NII's successor body, the Office for Nuclear Regulation, whether such a final report had even been written. Perhaps it was languishing in the files somewhere. After some weeks, and several promptings, he told me by e-mail that "I've been through our boxes and have not turned up a 'final report.'"

Boxes? What kind of an outfit is this?

So we don't know how close Sizewell A came to suffering Britain's worst nuclear disaster because the country's nuclear watchdog said one thing in its "preliminary" report and another to the media. Nor do we know why nobody was prosecuted for such an abysmal failure of critical safety systems designed to protect the public from radioactive fallout. It does not inspire confidence.

And, leaving aside the failure of regulation, it raises serious questions about how the owners of nuclear power stations now and in the future will go about their decommissioning. Was it party time at Sizewell once the station was off-line? Did the safety protocols get packed away ahead of time? Who exactly was in charge of the shut-down reactor? It suggests that the early phases of decommissioning may often be the most dangerous time for a power plant, and that the long saga of decommissioning the world's first generation of nuclear power stations is likely to be at least as error-strewn, deceitful, and dangerous as the operation of those plants.

░░░░

Decommissioning nuclear power stations is a Cinderella task. After all the fanfare of building the things and the massive amounts of power generated during operation, someone has to clean up the mess left behind. There are essentially three stages: removing the fuel; taking out the highly radioactive core and other contaminated steel, concrete, and other materials; and finally dismantling the plant and clearing the site. There are widely differing ideas about how quickly this could or should be done.

Some countries want to clear up the mess as quickly as possible and close the books. This also takes best advantage of in-house expertise about the plant and reduces the burden on future generations. It is, for example, what the US did at Maine Yankee near Portland, Maine, one of the country's first large commercial reactors, after it shut in 1996. It is what France, which has the largest fleet of nuclear power stations in Western Europe, has promised to do after a major shutdown of aging plants planned for the 2020s.

A second approach is to remove the fuel but then lock up the reactor for a few decades before dismantling and clearing the site. That is the approach favored by Britain, which plans to put reactors such as Sizewell A into what it calls "care and maintenance" for up to a century. In theory at least, that makes the dismantling easier and safer because some of the troublesome isotopes will decay away. It also postdates the decommissioning check.

A third option is to make the reactor safe and leave it intact indefinitely. In June 2011, the US government entombed two old weapons reactors at its Savannah River military complex in South Carolina in seven million cubic feet of concrete. Inside, managers stowed a time capsule containing a copy of *People* magazine with a report of the recent wedding of Prince William and Kate Middleton at Westminster Abbey in London. If anybody reads the report in less than 1,400 years, the entombment plan will have failed. The idea of this third option is to forestall further decommissioning costs, in effect forever. Whether future generations will thank Uncle Sam for this is another matter.[7]

However decommission is done, it is set to become a huge new global business for the twenty-first century. The UN's International Atomic Energy Agency reported in 2017 that there were already 160 large commercial power reactors permanently shut down in nineteen countries, with more than two-thirds in just four countries: the US, Britain, Germany, and Japan. Only seventeen had so far been dismantled and made permanently safe. The backlog is set to grow. There were another 449 reactors in operation round the world in 2017, including ninety-nine in the United States. Of those, 291 were more than thirty years old and scheduled to shut in the next decade—more if other governments join Germany in phasing out nuclear power.

Decommissioning a standard pressurized-water reactor (PWR) produces more than one hundred thousand tons of waste. That was what came out of Maine Yankee. Some sixty thousand tons of it was radioactive, including the steel reactor vessel, control rods, piping, and pumps.[8]

While the dismantling of Maine Yankee was complete in 2005, there is still nowhere to put much of the resulting waste. But at least Maine Yankee was dismantled. The decommissioning of American power plants shut more recently has mostly been shuffled off into the future, often as owners balk at the cost.[9]

Across the Atlantic, there is similar decommissioning disarray. A third of the European Union's power plants are likely to shut by 2025. The shortfall of cash set aside for decommissioning was estimated in 2017 at a staggering $135 billion.[10] Costs will rise. Decommissioning Germany's Soviet-designed power plant at Greifswald, on the Baltic coast, has produced more than half a million tons of radioactive waste, five times the PWR norm. The French company EDF's suggestion that its fifty-eight reactors can be shut for $23 billion was questioned by parliamentarians, who said it could cost three times as much.[11]

When the early power stations were being designed and built, little thought was given to how to dismantle them safely or cheaply. Often there are no surviving blueprints. The British Nuclear Decommissioning Authority (NDA), which is in overall charge of getting rid of that nation's fleet of shut power stations, reported that many "lack reliable design drawings. Many were one-off projects, built as experiments to test new approaches."[12] And few operators cared about the cleanup bill either. Decommissioners of the Connecticut Yankee power station found they had to remove 1.2 million cubic feet of soils contaminated by radioactive liquids that had been poured away. It must have seemed like a good idea at the time.

||||

As nuclear decommissioners balk at the task and put off the day, I wonder how the neighbors will respond to having the industry's dangerous detritus sitting over the back fence for decades to come. We may find out soon in Britain, where the first generation of commercial nuclear power plants occupy some of the nicest and most remote spots on the British coastline. None any longer operates and the power lines connecting them to the grid have ceased to hum with their energy. They aren't going away any time soon, however. Thanks to the British policy of pushing final decommissioning far into the future, these redundant hulks of the nuclear age will sit on sand dunes and scenic headlands, and nestle next to nature reserves and national parks, often into the twenty-second century. They will, moreover, probably be left unmanned. One engineer familiar with the industry's intentions told me, "The regulators haven't signed off on that, but the plan is for no on-site staff. Instead, the reactors would be fully instrumented with alarms to alert the local police, and five-yearly inspections." I am not so sure the neighbors will be happy with that.

I did a checklist of where Britain stands today. Sizewell A, the shutdown of which almost ended in disaster in 2007, is not due for site clearance until 2097. Until then it will remain, a looming presence next to the Minsmere bird reserve and the famous Aldeburgh concert venue. The year 2097 is also when Dungeness A in Kent should finally be removed from the end of Europe's largest shingle spit, home to a third of the plant species in Britain. Trawsfynydd power station is inside Snowdonia National Park in North Wales. To assuage aesthetic concerns when it was built, the design was handled by cathedral architect Basil Spence and the gardens were set out by landscape designer Sylvia Crowe. Trawsfynydd has been shut since 1991 and won't be demolished until 2083.

Bradwell sits on the Essex marshes east of London, right next to a pilgrimage site at Britain's second-oldest church, St. Peter-on-the-Wall, which was built by a missionary from the English island of Lindisfarne. The plant is locally notorious for the theft in 1966 of twenty uranium fuel rods from a store. A plant worker, Harold Sneath, figured they would be worth something for scrap. The theft was discovered only when a van driven by a friend of Sneath was stopped by police for defective steering.[13] Bradwell's cleanup date is 2092. We must hope that in the meantime future Sneaths don't try their hand at scavenging its contents.

The Berkeley power plant will dominate an officially designated site of special scientific interest (SSSI) on the banks of the Severn River until 2078. Nearby Hinkley Point A won't be gone until 2090. Hunterston, on a remote headland in southwest Scotland surrounded by the Portencross SSSI, is destined for demolition in 2081. On the Severn estuary, Oldbury has attracted 199 bird species to roost in lagoons at its water intake; it will linger until 2101. Wylfa, on the Welsh island of Anglesey, won't be gone until 2105, 134 years after it was completed. The last, a military reactor at Chapelcross in southwest Scotland, is not scheduled to be cleared until 2128.

Decommissioning these Magnox reactors is destined to be a difficult and expensive task. They are of a unique British design, so there is little foreign expertise to draw on. The current estimate is that each will eventually cost most than $2.6 billion, twice as much as standard PWRs. One reason for the price gap is that PWRs were built in factories and put together on site, so taking them apart again is relatively easy, whereas early British reactors are huge concrete civil engineering structures, built on site. They will be much harder to dismantle.

The nastiest legacy left by old Magnox reactors—as well as their successors, the fourteen Advanced Gas-Cooled Reactors (AGRs)—is radioactive graphite.[14] They used massive volumes of graphite to control the speed of nuclear reactions. Buckling graphite in Windscale's Pile No. 1 caused the fire there in 1957. Deteriorating graphite is the most usual reason for ending the life of British reactors, since it makes them unsafe and cannot be replaced without rebuilding the entire reactor. But the biggest problem with graphite arises during decommissioning.

During its lifetime in the reactor, graphite accumulates dangerous radioactive gases, most notably carbon-14, which has a half-life of nearly six thousand years. All told, Britain has some one hundred thousand tons of radioactive graphite sitting in its reactors, almost half the world total. Unless scientists can find ways of collecting the gases built up inside the graphite, it will eventually make up a third of Britain's most dangerous "high-level" radioactive waste.[15]

The graphite problem is one that Britain is going to have to solve for itself. But the more general issue of what to do with old radioactive nuclear power stations is one that is going to become of increasing public concern in the United States, which has the world's largest inventory of aging plants, and around the world in the decades to come. Exactly how long will we be willing to tolerate the existence of these radioactive relics in our landscapes when they are not even generating power?

Meanwhile, back in Britain, the decommissioning problem that really keeps the nuclear bosses awake at night is at Sellafield, which for more than half a century has been Britain's biggest source of radioactive waste and the place where all the waste from elsewhere eventually ends up. It is Europe's largest nuclear dustbin.

Chapter 22

Sellafield

*Stone Circles and
Nuclear Legacies*

Not many people visit the Grey Croft stone circle in western Cumbria, in Britain's far north. That's a shame. The ten chest-high stones set in green pastures have been commanding views toward the mountains of the Lake District and over the waters of the Irish Sea for five thousand years. They are still worth seeing, but the sense of timelessness is rather spoiled these days. For the vista is now dominated by something else. A few hundred yards from the stones sits Britain's largest industrial site, the Sellafield nuclear complex. If the stones left by a long-forgotten Neolithic civilization still invite awe, the twenty-first-century nuclear henge often invites fear. And the task of making this corner of Cumbria safe for future civilizations is daunting.

After admiring the stones for a few minutes, I walked down the hill, across a field, over a small stream, and followed the double line of high mesh fences topped with razor wire that surrounds Sellafield. I was soon stopped by a squad car of the Civil Nuclear Constabulary, a dedicated police force protecting British nuclear installations. The two constables inside carried assault rifles. Nobody is taking chances.

Sellafield is Britain's brooding nuclear nightmare. Packed onto its 2.3 square miles are tanks, pools, silos, and outsize buildings containing the lethal remains of Cold War bomb making, the accumulated radioactive waste from sixty years of nuclear electricity generation and, for good measure, the world's largest stockpile of terrorist-ready plutonium. There are open-air ponds full of decades-old corroded nuclear fuel, tanks of hot liquid wastes from nuclear reprocessing, and potentially explosive remains inside the sealed sarcophagus from the fire in 1957. Nowhere else on the planet—not even the Mayak complex in Siberia—contains so much radioactive material in such a confined area. A terrorist attack, earthquake, or cataclysmic accident at Sellafield could destroy the entire place and lay waste to swathes of northern England.

Having satisfied the nuclear police, I walked on a short distance down a quiet country lane, away from the plant to Ponsonby church, part of an estate that William the Conqueror gave to the Ponsonby family more than nine hundred years ago. Oddly enough, a scion of that family is a neighbor of mine in London. Lambs cavorted around the churchyard in the bright sun, as they must have done each spring for centuries. Inside the church, a notice warned that if the Sellafield siren blew, warning of a radioactive release, nobody should leave the church. The doors should be shut and news sought by tuning into a radio.

There have been three phases in the life of Sellafield. During the 1950s, when it was still called Windscale, it made plutonium for British bombs. Later, it became the center for handling waste from Britain's growing civil program for generating nuclear power. It converted some of that waste into plutonium that nuclear engineers hoped could become an endless source of fuel for a hoped-for "plutonium economy." In the twenty-first century, with those ambitions lying in ruins, it is retooling for its final and longest task: shutting itself down. So, after years under the management of the quasimilitary Atomic Energy Authority and quasicommercial British Nuclear Fuels, the new landlord here is the Nuclear Decommissioning Authority.

The NDA has a formidable job. In the early days of operations at the old Windscale, few records were kept of building designs or waste inventories. As the government's own website puts it, back then "in a heady atmosphere of scientific discovery, plans for future dismantling were barely considered." That disregard for the future has proved expensive. Sellafield currently spends about $4 billion a year on cleanup. The estimated cost of clearing the site by 2120 keeps on rising. At the time of writing it was $153 billion, almost three-quarters of the total British nuclear cleanup bill of $210 billion.[1]

My earlier tour around the hinterland of Sellafield, among the radioactive salt marshes and lost seaside resorts, had been troubling. But most people I spoke to on the outside told me that the real radiological hazards still lay inside the fences. Sellafield harbors a toxic legacy of waste from its past endeavors that is, if anything, more dangerous than ever. My tour confirmed that view. Many of the buildings that hold the stuff now crack, leak, corrode, sprout weeds, and accumulate dark radioactive sludge. I found that Sellafield's engineers, in their macho way, were proud of the toxic legacy that they superintended—including, they almost boasted, the two most hazardous industrial buildings in Western Europe. But I found a second legacy: an equally corrosive legacy of duplicity, secrecy, and plain lies. It goes back at least as far as the toxins, and it may be just as hard to clean up. It is a legacy common to most nuclear activities, and particularly in those countries where bomb making has been part of their remit. From Hanford and Rocky Flats to Ozersk and the rest of the old Soviet gulags, the story is frighteningly similar. The legacy of what the historian Kate Brown calls "Plutopia" is disturbingly universal.

Altogether, Sellafield has 240 radioactive buildings awaiting decommissioning. The most immediately obvious landmark, with its towering chimney, is the plutonium-producing Windscale Pile No. 1 that caught fire sixty years ago. After the fire, they sealed it up as best they could, and nobody has yet dared breach the seal. Inside, the buckled graphite core still holds the Wigner energy that operators were trying to remove on the day of the fire. Disturbing the remains could cause the core and its estimated 16.5 tons of uranium fuel to catch fire again. Decommissioning has long been promised. Former company secretary Harold Bolter, in his 1997 book *Inside Sellafield*, said it would be done by the end of the century. He meant the twentieth century. Then it was promised for completion by 2005, and 2015. In 2017, it still waited. In the decommissioning pecking order, it has fallen behind four other buildings that the NDA says are even more dangerous.[2]

The deadly four each hark back to the same era as the piles. They contain fuel and waste that should have been made safe decades ago. But as energy minister Chris Huhne put it in 2011, "when waste started piling up, we crossed our fingers and hoped it would go away." It didn't. Each of the deadly four will take billions of pounds to make safe.[3]

Take Building 29. It contains a pond 330 feet long that has six times the volume of a typical Olympic swimming pool. The pond received the spent fuel rods as they were pushed from the backs of the piles. They stayed there to cool before the fuel was removed from its cans and sent for reprocessing to extract their precious plutonium. After the fire, the pool was retired. It was hurriedly taken out of mothballs in the early 1970s as an emergency store for spent fuel delivered to Windscale during a strike by coal miners. Nuclear stations were run at full pelt then to make up for the loss of electricity from coal power stations. Windscale's reprocessing line couldn't keep up. Skips of spent fuel coming by train daily from power stations across the country were dumped into B29.[4]

The backlog grew. After a few months, the fuel began to corrode. This added to the radioactivity in the pond, and after a time meant that the fuel could not be reprocessed at all. So the fuel was simply left in the pond. There it remained, continuing to corrode. By 2008, an impenetrable layer of radioactive algae, corrosion products, and broken bits of fuel had formed on the bottom of the pond. That year, after decades of delay, work began to clean it up. It has proved slow going. It took eight years before the first pounds of sludge were scooped into a drum and mixed with cement.[5] It will take several years more to package the fuel and sludge and put it all into a nearby store. Then they must remove the radioactive water from the pond. That won't start until 2019 and will take a further ten years.[6]

When B29 was closed for the first time, in 1957, it was replaced by B30—another giant pond, this time five hundred feet long. Like its older brother, it is open to the elements. Until 1985, it received spent fuel from Sellafield's Calder Hall reactors and the new generation of power-producing Magnox reactors around the country. The fuel was meant to stay there for a few months before reprocessing. But like B29, when things weren't going smoothly at the reprocessing plant, the fuel stayed too long and began to corrode. Again, there it remains.

B30 has an even thicker layer of muck than B29, including more than a ton of plutonium. Sellafield managers don't show B30 to visitors, but pictures leaked in 2014 revealed weeds growing around the pond and radioactive algae floating on the water. Sellafield workers refer to it as "dirty 30" because it has long been the site's biggest source of occupational exposure to radiation. Workers can remain in some areas around the pool for only two or three minutes.

Close to these two radioactive ponds are two giant silos built to store the highly radioactive cladding cut off spent fuel from the piles and Magnox reactors before it went for reprocessing. B41 was built in 1950 and contains six giant hoppers, rather like grain hoppers, each seventy feet high.

In 1965, with its hoppers full to the brim, it was closed. Plans for emptying the aging hoppers during the 1990s came to nothing. Around then, it was discovered that as the contents corroded, they generated hydrogen that might catch fire. So the guardians of the silo began to pump in argon gas, which would stifle any conflagration. In late 2016, they completed fitting giant stainless-steel doors to the hoppers. The following year, one of the hoppers was cut open for the first time in a half-century, allowing the contents to be safely removed, a process that might start as soon as 2020.[7]

B41 also had a replacement when it shut: B38. To prevent any risk of the cladding catching fire, designers decided to store it under water rather than in hoppers. In busy times, B38 also stored skips of fuel. But at some point, the pool bottom cracked and radioactive sludge began to seep into the soil. Nobody noticed until, in 1976, construction workers on another project nearby found they were excavating a radioactive marsh. Tens of thousands of gallons of sludge rich in cesium-137 had escaped.[8] There it remains. Managers call B38 and B30 the two most dangerous industrial buildings in Western Europe.

Besides the "big four," there are numerous other toxic time bombs ticking away at Sellafield. Many are long-standing. Take B701, which stored tanks of highly acidic and extremely radioactive liquid waste from reprocessing. It was abandoned after the 1957 fire and supposedly emptied. Twenty-two years later, however, construction workers nearby discovered a massive leak, which they tracked back to pipework in B701. The forgotten building turned out to be so radioactive that nobody could safely enter. Using video cameras, inspectors concluded that 2,600 gallons of some of the most dangerous liquids on the site had sloshed into the soil. More than one hundred thousand curies were on the loose, more than twice as many as in the airborne release from the Windscale fire. And there, by all accounts, they remain to this day.[9]

The combative chairman of Parliament's National Audit Office, Margaret Hodge, was not impressed by this litany of problems when she investigated in 2012. She said that the ponds and silos posed an "intolerable risk to people and the environment."[10] Nobody contradicted her. While the first belated steps to empty them have begun, timetables have been slipping. In early 2016, the NDA postponed the completion dates for emptying B29 and B41 by five years—to 2030 and 2029, respectively. Even those deadlines may prove hard to meet. "We have to do a lot of R&D just to characterize the inventory, before we can work out how to retrieve the materials," Paul Howarth, the managing director of the National Nuclear Laboratory, a government body based at Sellafield, told me.

Political vacillation doesn't help. In 2008, most Sellafield activities were privatized. The aim, said the energy minister at the time, Mike O'Brien, was to "get to grips with the legacy after decades of inaction." But as costs ballooned, O'Brien's successors decided in early 2015 to rationalize everything again. Whoever is in charge, the predicted Sellafield cleanup bill keeps going up, from $650 million in 1980 to $103 billion in 2014 and $153 billion in 2016.

IIII

While much of the danger at Sellafield lurks in abandoned equipment and buildings, there are also continuing stockpiles of waste that are at least as troublesome. One source of concern is the acidic and highly radioactive liquids that continue to be generated by Sellafield's two reprocessing plants. In theory, this stuff is supposed to be quickly treated and stabilized. This is to eliminate the risk of an accident of the kind that happened at Mayak in Siberia in 1957, when tanks of just such waste boiled and exploded. But whatever the theory, in practice there is a big backlog of this nasty stuff, sitting in giant stainless-steel tanks, each twice the size of a shipping container.

The tanks require constant cooling, and a failure of the cooling system would be disastrous. According to a Royal Commission report on Britain's nuclear industry in 1976, known as the Flowers Report, it would "cause the solution to boil dry and the heat generated would then disseminate volatile materials to the atmosphere and cause widespread contamination."[11] More recently, Gordon Thompson, of the Institute for Resource and Security Studies in Cambridge, Massachusetts, has warned that if Sellafield experienced such an accident, "a large area of land could be rendered unusable for a period of decades. Neighboring countries could be significantly affected."[12]

Such concerns led regulators at the NII in 2001 to order Sellafield's managers to cut stocks of the liquid waste by almost 90 percent by 2015— from 416,000 gallons to no more than 53,000 gallons. That required speeding up the existing system for making the liquids safe, by first evaporating them to reduce the volume and then converting them into a solid glasslike material, a process called "vitrification." Yet by the start of 2016, with the target date expired, those managers were less than halfway there. They blamed a seven-year delay in finishing their new $840 million evaporator and repeated shutdowns at the vitrifier. The NII's successor, the Office for Nuclear Regulation, might have demanded that reprocessing be suspended until there was adequate capacity to handle the resulting liquids. But instead, it simply abandoned the target.[13]

It is hard to see a safety case for such leniency. One of the concerns has always been that an earthquake could breach the tanks containing the highly radioactive liquids or disrupt the constant cooling that keeps them safe. After the earthquake and tsunami at Fukushima, British regulators increased their assessment of the seismic threat to Sellafield, extending the zone around Sellafield covered by evacuation plans from 1.2 to 3.7 miles. So their assessment of the risks posed by the large stores of hot liquids must also have increased. The new regulator appears to have decided that Sellafield's operational convenience takes priority over risk assessment.

▌▌▌▌

In the longer run, Sellafield's lethal legacies have to be found a final resting place. For much of the waste, that ultimately means burial deep underground. A couple of miles from Sellafield's back fence, within yards of the Lake District National Park, is the burial site the authorities keep coming back to. During the 1990s, they proposed that Longlands Farm, just down the road from the Ponsonby church, should be the entry point for a test repository that might one day extend for eight square miles underground. After a damning planning inspectors' warning that the porous local limestone rocks could transmit radioactive leaks into water supplies, the government threw the plan out. A decade on, new ministers revived it, only for the Cumbria County Council to vote against it in 2013.

In 2015, I met the former county leader, Eddie Martin, at his home a few miles from the plant. He told me he thought the government would persist. "They want the disposal facility here not because the geology is favorable—it isn't, as the old miners know it is riddled with faults and fissures—but because they think we won't object."

The people of West Cumbria have never really had a choice about whether they want Sellafield. The first phase of military operations began in great secrecy. By the time they knew there was a plutonium factory in their midst, they had become dependent on it for jobs and tax revenues. At the time, the bargain didn't look a bad one. In the boom times when nuclear power was a growing industry and Sellafield was the reprocessing hub, the complex provided most of the area's jobs, spent a lot of money, and had its tentacles in all manner of local social activities. Even the local Member of Parliament from 2005 to 2017, Jamie Reed, had been a PR man at Sellafield and was sponsored in Parliament by the GMB, the largest blue-collar union at the site. When he resigned as an MP, it was to take another job there. But even if the boom years felt good, a technocratic decommissioning agency overseeing the closure of the place is unlikely to attract such loyalty.

There was disgruntlement now, said Martin. The people of West Cumbria felt they had been taken advantage of: "Sellafield has stored the country's nuclear waste and operated some of its most dangerous plants for almost seventy years. For what we have done for the country, the streets should be paved with gold. Instead, we have been largely ignored. Our infrastructure is poor. Too many children live in poverty. And in the back of our mind, we know that if there is an accident the whole area could become uninhabitable."

Chapter 23

Hanford

Decommissioning an Industry

May 9, 2017, was just another day in the life of the defunct Hanford nuclear reservation, a remote part of the state of Washington covering some six hundred square miles of sagebrush desert. Then alarms sounded. Almost two thousand site workers were told to take cover indoors, and aircraft were banned from flying over the reservation for several hours. The roof of a tunnel had collapsed, exposing rail cars that had been loaded with radioactive waste from plutonium production and shunted underground for burial decades before.[1] Suddenly, Hanford was back in the news.

Straddling the Columbia River, the reservation was America's primary bomb-making factory for more than four decades. It was where they made the plutonium. At peak production, during the 1960s, its nine reactors went through seven thousand tons of uranium fuel annually. Its five reprocessing plants turned that spent fuel into more than four tons of plutonium a year, enough to make 650 Nagasaki-size bombs. They eventually delivered sixty-seven tons of the metal for the American arsenal, before business halted with the end of the Cold War in 1989.

It was a huge task, requiring much of the electricity produced at the giant Grand Coulee Dam upstream on the Columbia, the largest power producer in America.[2] And now the legacy of contamination is equally mind-boggling. Since production ceased, Hanford has been conducting the country's largest-ever environmental cleanup program. The current cost is $2.3 billion a year. By the time it is done the bill will be more than $100 billion—probably much more.

The site holds an estimated twenty-five million cubic feet of solid radioactive waste, enough to fill London's Royal Albert Hall eight times over. Much of it is buried in forty miles of trenches and tunnels, up to twenty feet deep and fifteen feet wide. "They are jammed with God knows what," says Tom Carpenter, founder and director of Hanford Challenge, an NGO monitoring the site. "There are few records. But we know some of the trenches are full of radioactive animal carcasses from animal experiments. They include eighteen alligators." Two tunnels contain railway tracks that allowed entire trains of railroad cars full of radioactive waste to be shunted in and covered over: thirty-six cars are reportedly down there. It was one such tunnel, running directly out of the main reprocessing plant, the thousand-foot-long PUREX plant, that caved in along a twenty-foot stretch in 2017.

Elsewhere on the reservation there are huge stores of spent fuel. One building contains six cooling ponds, each the size of an Olympic swimming pool, holding 120 million curies of strontium and cesium waste— nearly as much as was released by the Chernobyl accident. But the headline problem is the huge volumes of acidic and radioactive reprocessing waste accumulated over the years. All told there are fifty-six million gallons of these liquids and sludges, stored in 177 tanks up to seventy-five feet in diameter. That is around 30 percent of all the most radioactive liquid wastes being held at reprocessing plants around the world. The total radioactivity in the tanks is an estimated 350 million curies, which is more than twice the emissions from the Chernobyl accident.

The Hanford tanks have been seeping into the ground almost from the start. That fact was a state secret until the late 1980s. The sixty oldest tanks are the worst. They have single shells and have leaked around a million gallons into the soil. By 2017, two of the newer doubled-shelled tanks, each holding almost a million gallons, were also leaking.[3] Much of this contamination is slowly flowing in underground streams toward the Columbia River.

Protecting the river from the radioactive pollution is the number one cleanup priority, says the Department of Energy, which is in charge of the reservation. It says it wants to protect some of the river's best salmon-spawning habitat, which is in the Hanford reach running for fifty miles past the reservation. But keeping the eighty square miles of underground water flowing through the site from reaching the river requires a constant program of pumping up water and sending it for treatment. So far only a small fraction has been treated.

A better idea is to stop the leaks in the first place by finding a safe disposal route for the liquids. The current aim is to remove the most dangerous liquids from the tanks and heat them with glass-forming materials to create solid blocks that could one day be buried—maybe at Yucca Mountain in Nevada, the proposed waste dump for the nation's most dangerous radioactive waste. Work constructing the giant "vitrification" plant began in 2002. At the time of writing, the expected completion date is twenty-five years behind schedule. Operations are not set to begin until 2036 and, once under way, would take forty years. At $17 billion and counting, the project is also way over budget.[4]

Many, including former plant engineers who have turned whistle-blowers, believe the design of the vitrification plant is not fit for the job and should be abandoned. They believe there is a serious risk of particles of plutonium settling out in the plant processing tanks, creating the potential for a criticality accident with a big release of radiation. One of the whistle-blowers is Walt Tamosaitis, a project troubleshooter who was fired when he refused to withdraw his warnings. He won a court settlement for unfair dismissal but, says Carpenter, his concerns remain largely unaddressed.[5]

The delays over treating the tank waste are symptomatic of slow progress and big spending at Hanford. After almost three decades, little of the waste and few of the tanks or buildings have been cleaned up. Far away in Washington, DC, some question the continuing cost, arguing that Hanford has become a "money sink" with ever more spent to achieve ever less.[6] It seems to some like a twenty-first-century pork barrel. "A lot of money is being spent to mitigate risk that is very small," one policy advisor told author Andrew Blowers in 2013.[7] Perhaps, some critics say, it would be better to put up a fence and walk away.

Local environmentalists are scandalized. "We have got to clean up the site," Dan Serres, of the NGO Columbia Riverkeeper, told me at his office in Seattle. The tanks should be emptied and the trenches dug up. "The area is too attractive, for fishing and other things, to leave behind a hazard," he says. "In a hundred years, I'd hope the Native Americans have their treaty rights to this land restored," agreed Chuck Johnson, of Physicians for Social Responsibility. "They should be able to eat the fruits of the land, drink the water, and eat the salmon."

But will they? Carpenter, who sits on an advisory board at Hanford, is not so sure. "You are never going to dig all the waste there up," he told me. The tanks will have to be dealt with, but "most of Hanford's waste volume-wise is going to stay put. Hanford is going to be a national sacrifice zone for hundreds of years."

Hopefully, that is not the fate that awaits Andreeva Bay.

||||

Andreeva Bay is an isolated spot—at least for Russians. It is somewhere past Murmansk on the Kola Peninsula going north, where Russia meets the Arctic Ocean. But for Norwegians it feels a lot closer. Andreeva Bay is only twenty-five miles from the border. That is why Norway is none too happy that the Russian Northern Fleet has for decades been using the bay's frozen foreshore to store the spent fuel from reactors that once powered around a hundred submarines.

The bay's submarine base shut in 1992, after the collapse of the Soviet Union. But the departing fleet left behind on the quayside twenty-two thousand packages of spent fuel from submarine reactors. Some of the spent fuel was left in large open-air ponds that cracked and leaked, before being belatedly moved to "temporary" dry stores. There they remained. After years of pressure from Norway there is now at last a plan to move this nasty legacy. The task will cost hundreds of millions of dollars, paid for by the Norwegians, the British, and others, and has involved lots of new infrastructure. First there was a sarcophagus built over the waste stores so the spent fuel could be packed safely into casks for transportation. Next, the Italians built a special ship, the *Rossita*, to get the spent fuel as far as Murmansk. From there it will be loaded onto specially made rail cars for the journey to Russia's main reprocessing and waste management center, the Mayak complex, more than 1,300 miles away at Ozersk, in the Urals. The first shipment left the port in June 2017.[8]

That is not the end of the cleanup, however. The ponds that first housed the spent fuel have been leaking water contaminated with cesium-137 into the soil. The radioactivity has been flowing toward the ocean. How far it has gotten remains unclear, because in 2014 the Russian navy banned Norwegian marine biologists from accessing the site to map the contamination.[9] But by early 2017 more than thirty-five thousand cubic feet of radioactive soils had been dug up to try and stanch the flow.

Whatever their failings, we should probably be thankful for the tanks and silos at Andreev Bay. They contain what the Northern Fleet never got around to dumping at sea. Between 1959 and 1991, the fleet dropped to the floor of the Arctic Ocean at least six submarine reactors still containing their spent fuel. It emerged in 2011 that the sea bed in the scuttling zone also holds nineteen ships and seventeen thousand containers holding radioactive waste, eight more nuclear reactors without fuel, and an entire nuclear submarine with two reactors that still have their nuclear fuel on board.

This submarine is the K-27. It suffered a major "reactor malfunction" on its third voyage in 1968. Radiation began leaking. Nine seamen died of acute radiation poisoning after trying to fix the problem. Many other crew members became sick. Eventually, the Soviet authorities decided to fill the reactors with bitumen to prevent further leakage, after which they sank the vessel at a depth of just one hundred feet.[10] Bellona, a Norwegian NGO, has campaigned for two decades for the K-27 to be brought back to the surface. Its nuclear specialist, Thomas Nilsen, claims that "radiation leakages will come sooner or later, if we just leave it there." The sub, he says, "was rusty even before it was sunk. Leakages of radioactivity under water are nearly impossible to clean up."[11]

The Northern Fleet has suffered a litany of accidents and been guilty of much reckless waste disposal over the years. In 1989, its K-278 submarine sank in the Barents Sea after a fire. It went down with forty-two crew members, a fully fueled reactor, and two torpedoes carrying nuclear warheads. A 1993 report to the UN London Dumping Convention warned that "active warhead materials were in contact with seawater" and predicted that plutonium could escape into important fishing waters, creating "a zone of contamination which will be both highly active and chemically toxic." Attempts to seal the torpedoes and keep the warheads intact were reportedly partly successful but might not last beyond 2025.[12]

The US has dismantled more than a hundred nuclear submarines; their reactors are lined up at Hanford awaiting burial in the reservation's trenches. But Russia has many more nuclear submarines moored offshore waiting to have their reactors removed. Most of the reactors will eventually go into storage in a new complex of giant tanks being built at Saida Bay, outside Murmansk.[13]

Meanwhile, Britain has its own version of the Russian problem. Twelve of its nuclear submarines are sitting in what officials call "afloat storage" at the Devonport naval dockyard in southwest England, where they are disturbingly close to houses and primary schools. The subs have their reactors intact. Most have their nuclear fuel, too, after government nuclear inspectors ruled in 2002 that safety facilities at the dockyard were not good enough to allow safe removal of the fuel.[14] A further seven British submarines are at the Rosyth dockyard in Scotland, though they have all had their fuel removed.[15]

The original plan had been to sink all of Britain's submarines at sea, like the Russians. Even after Britain stopped dumping radioactive waste at sea in 1983, the Ministry of Defence insisted the scuttling should go ahead. It only accepted that the rules should apply to military waste a decade later. Where the redundant British submarines should now go has yet to be resolved.[16]

The fate of nuclear submarines remains a relatively small, if high-profile, part of the overall problem of what to do with radioactive waste. As we will see in the next two chapters, making this waste safe and disposing of it in a publicly acceptable manner is probably the biggest and most politically fraught part of the nuclear legacy. We start in Germany, where the generation of nuclear power is drawing to a close, but where, thanks to decades of mismanagement of radioactive waste, the country's nuclear nightmare may be far from over.

Chapter 24

Gorleben

Passport to a
Non-Nuclear Future?

Helmet and safety boots on; dosimeters clipped to white overalls; flashlights in pockets; breathing apparatus in shoulder bags. With all the kit checked, we dropped in a metal cage for 1,600 feet down a shaft. A car was waiting at the bottom to drive us another 800 feet down a subterranean roadway toward the bottom of the mine. The Asse salt mine in Lower Saxony was warm, dry, and fresh, thanks to massive fans sending air from the surface to keep a workforce of 120 people safe. But what were they all doing? For not a teaspoon of salt had been sent to the surface since mining stopped half a century ago.

Instead, deep beneath the forests of northern Germany, an expensive nuclear farrago is being playing out. Here, sealed in caverns in the salt, lie drums containing tens of thousands of tons of radioactive waste buried during the 1970s. Forty years on, there are fears that the mine could be overwhelmed by water flowing in from the rocks that surround it. That water could wash radioactivity to the surface. Nuclear-averse Germans do not want that, or to inflict the prospect on future generations. The workforce has gone back underground to figure out how to extricate the drums from their caverns and bring them back to the surface. It is a task that looks set to take decades and cost billions of dollars. And there are those who say the solution will be worse than the problem.

The radioactivity locked up in the waste in the caverns is not huge. This is not spent nuclear fuel. It is all designated as either "intermediate" or "low-level" waste. Yet the scheme to dig it up is threatening to further undermine public confidence in German nuclear engineers, just as the country attempts to dismantle its fleet of nuclear power stations and embark on a nuclear-free future. Bemused Germans are asking, if "intermediate" nuclear waste that was declared safely buried thirty years ago can so quickly become unsafe, how can they trust schemes to bury much more dangerous waste that needs to be kept out of harm's way for thousands of years? Welcome to nuclear mismanagement German-style.

The Asse mine is near Wolfenbüttel, a town close to the Oker River known for its timber-framed houses. The mine shut in 1964 after fifty-six years of the extraction of salt deposits. The following year, the German government's research ministry bought the abandoned mine, ostensibly to investigate its suitability for disposing of radioactive waste. But after two years, without waiting for any research findings or making any public announcement, ministry officials quietly turned the mine into what amounted to a permanent dump. Radioactive waste from the country's power stations, medical and military facilities, and industry began to be delivered here and sealed into chambers. According to a later report from the Federal Office for Radiation Protection (BfS), "Almost all the low- and medium-level radioactive waste that was produced in West Germany between 1967 and 1978 was deposited in Asse . . . under the cover of research."[1]

The volume is astonishing, certainly for "research." In all, they buried 125,787 drums in thirteen chambers. Each iron drum holds fifty-two gallons of waste, making enough waste to fill twenty Olympic swimming pools. Again oddly for a research project, the precise contents of the drums were never logged. Estimates of the plutonium content, which will remain radioactive for thousands of years, have risen to around sixty pounds, though nobody really knows.

Despite its worrying contents, the mine has some homely touches. On a walking tour of its tunnels with Ingo Bautz of the BfS, who oversees activities at the mine today, I turned one corner to find a small, backlit statue of the Virgin Mary sunk into a carved-out hole in the salt. Further on, surrounded by scaffolding and huge ventilator ducts, was a neat office desk with a black Anglepoise lamp, propped on wood blocks to prevent its metal feet gouging the salt.

Those who buried the waste thought they had put it safely out of harm's way, said Bautz. But that view has changed. Even while the drums were being lowered in, the mine's excavated chambers were buckling under the weight of the surrounding rocks. As the walls bent, cracks formed. "The mountain has been moving," Bautz said. He took from his pocket a photograph taken in the mine in 1967 that already showed collapsed walls. Water began to seep through the cracks into the mine in 1988. "There were people who said all along it wasn't a good idea to put radioactive waste down here," he said. "Everyone knew there was a big risk of flooding. Other local salt mines had flooded in the 1960s. But the risks were ignored."

Some engineers I spoke to believe there would be no harm if the mine did flood. Any radioactivity liberated from the drums would take centuries to reach the surface, they said. But there was certainly no public debate on the matter. As public concern grew in Germany about radioactive waste, there was official silence—about the secret burials and about the water seeping into the mine.

It was only in 2008 that the scandal became public knowledge. Since then, new managers have been backfilling much of the mine to stop further buckling and to block potential routes for escaping floodwaters. But the slow collapse continues. Water enters now at a rate of about three thousand gallons a day, said Bautz. As we walked on, I heard the sound of water and saw a large tank where that water is collected. In a large mine, the amounts of water were tiny, and while the water is salty, it is not radioactive. But public concern about the mine is now so great that engineers are pumping the water to the surface and down pipes for seventy miles before pouring it down another salt mine.

After walking and driving further downward, stopping at gates in the road, and responding to mysterious sirens and flashing red lights, we reached the bottom of the mine. "This is the oldest level. These walls are only good for a couple of years," said Bautz. Further collapse and a massive invasion of water could engulf the entire mine, he warned, dissolving its salt walls and pouring into the chambers below that hold the radioactive waste. Then the drums would corrode and the radioactivity would surge to the surface. That's the nightmare scenario. That is why, in 2011, the BfS ruled that the only solution was to bring the drums out first and find somewhere else to bury them.

Many dismiss the nightmare scenario as scaremongering. Peter Ward, a British nuclear engineer who has worked on radioactive waste disposal in Germany for many years, told me, "The most dramatic estimates, including the failure of all barriers, predict a minimal release at the surface in forty thousand years. Practically all of my colleagues agree that retrieving the waste from the Asse is stupid, unpractical, and a waste of resources. It is also extremely dangerous for the work force," because of the ever-present risk of the mine collapsing. Ward said a 2011 technical review that considered the options recommended leaving the waste in the ground. But BfS made a "political" decision to dig it all up.

Like the original burial, the decision to retrieve the drums may have seemed like a good idea at the time. But even those who supported the decision concede that it is proving much harder than anyone imagined. Just checking the state of the sealed chambers that contain the drums is painfully slow work. Engineers drilling through the sixty-five feet of salt to the chambers don't know what they will find and cannot risk a release of radioactivity that would swiftly reach the surface via the mine's ventilation system. Moreover, the drilling itself has to be done in slow motion because a spark could ignite a fire that would engulf the drums and spray their contents through the mine. An explosion inside drums of intermediate-level waste down a salt mine in New Mexico in 2014 took three years and $2 billion to clean up. Anxious not to repeat that, the drilling teams at Asse completed just one borehole into one of the thirteen chambers in their first four years.

Once the drums are released from their chambers, taking them to the surface will require reengineering the entire mine, including sinking a second shaft and opening up new galleries, said Bautz. "Sinking a new shaft is extremely dangerous," said Ward. Bautz conceded that "we can't rule out that the mine could flood. If that happened, retrieval of the drums would be impossible. We would backfill it all." That, at least, he and Ward agree on.

What are we to make of this? "It is a disastrous situation," said federal environment secretary Jochen Flasbarth when I met him in Berlin the following day. All this activity, all these people, all these years, and all they had to show for it was a single borehole into a single chamber. Who is to blame? Is it the cavalier foolhardiness of past engineers or the extreme risk aversion of today's public and politicians? The answer is both. German policymaking has flipped between the two extremes. For a country that prides itself on rationality and pragmatism, little of either seems to have been on display in Asse.

But whoever is to blame, and whatever the real risks, the price tag is staggering. It is already costing $160 million a year simply to keep the mine in a safe enough state for work to continue on preparing for retrieval. No drums are expected to leave the mine before 2033 at the earliest. The mess at Asse is set to take decades and upward of $4 billion to resolve. That is a lot of time and money when the best solution might still be to close up and walk away.

IIII

The debacle at Asse is one small element in Germany's growing radioactive waste mountain. The country is the first major nuclear country to pull out of the business altogether, with all its nuclear power plants to be shut by 2022. So, now that their victory has been secured, I was keen to find out how antinuclear campaigners were considering this legacy of waste. Did they accept that Germany had to find ways of handling the stuff and making it safe? Or did their blanket opposition to all things nuclear leave them as perpetual critics of whatever plan the authorities came up with? If so, might they not end up prolonging their country's nuclear nightmare?

I began my pursuit of answers to these questions by meeting one of the country's most venerated and certainly most steadfast nuclear opponents. Wolfgang Ehmke cut a startling figure in the snow of a January morning in the snug little German village of Gorleben, set amid forests near the Elbe River. He was proud to have been among the first antinuclear Green Party activists. He decided forty years ago to dedicate his life to preventing this village from being turned into the hub of the Germany nuclear industry, to thwart the government's ambition to build a plutonium reprocessing plant at the surface and a waste burial site in the salt dome beneath.

Balding now, with yellow drainpipe trousers and seemingly impervious to the cold, he had stuck it out. Even after German chancellor Angela Merkel announced in 2011 that she would shut all the country's nuclear power plants, he did not feel able to desert his post. The village already housed an "interim store," where more than a hundred flasks of Germany's most radioactive waste were stacked up to the roof, and nuclear engineers were busy measuring up the salt for their disposal. The government denied any decisions had been taken, but Ehmke and the other protesters were in no doubt that waste generated by Merkel's nuclear shutdown could still end up buried permanently in the Gorleben salt unless they stopped it.

The motivations of the activists vary. For many, the cause is at root about politics: their sense that nuclear power is indissolubly linked to nuclear weapons. To others the risks of the technology getting out of control in some accident are unbearable, and they believe that keeping it safe would require infringements of civil liberties that they find equally unacceptable. Still others believe that radiation is far more dangerous than officially recognized. Finally, many argue that the technology, particularly of waste management, creates risks and responsibilities for future generations that are fundamentally immoral.

On the cold, snowy morning of my visit, as on every Sunday for forty years, a group of activists held prayers in the forest within sight of the waste store. They asked for "deliverance" from the nuclear threat. Beneath four wooden crosses, a banner read "*Wachet und betet*"—Watch and pray. Ehmke, who told me four generations of his family had joined him on the barricades, had a different, more secular, language: "The fight goes on; our resistance must never be broken," he said.

||||

The problem for Germany is that the cathartic decision to phase out nuclear power generation does not draw a line under its nuclear problems. The waste left behind must still be dealt with and, like other nuclear nations, Germany has never figured out what to do with it. That waste awaiting a final home includes the drums to be dug up from Asse; the vast volumes of radioactive rubble that will be created as the pensioned-off nuclear power plants are dismantled; the thousand or more flasks of spent fuel currently in the interim store at Gorleben and at power stations round the country; and the waste waiting to return from reprocessing plants in France and Britain. The most politically charged question is where to put the most dangerous "high-level" waste. The answer for many is simple and the same as it was forty years ago: the salt mine at Gorleben.

It looks like a no-brainer. The planned Gorleben reprocessing plant was never built, but the interim store created there contains 113 flasks of high-level waste, mostly the products of the reprocessing of spent German reactor fuel in France and the UK. And below ground, there is a vast network of caverns that scientists have been working on for years, preparing them for becoming the country's nuclear burial ground. The place is officially still a research facility, but so was Asse. What would be simpler than choosing the mine across the road from the interim store for the final burial?

Gorleben has been the most potent focus of Germany's antinuclear movement for more than a generation. In 1979, with Three Mile Island fresh in many minds, some one hundred thousand protesters gathered there. In 1984, the Greens set up a permanent camp. They blockaded trains bringing waste to the interim store. Farmers dumped manure and potatoes and cemented themselves to the tracks. Many people lived at the camp until police demolished it.[2]

A mythology has built up. Ehmke and other activists renamed the local area the Free Republic of Wendland, the ancient name for the territory of a long-disappeared Slavic tribe. Wendland had its own radio station, flag, and "passport." Not everyone in Wendland is antinuclear, of course. Many would rather like to host the nation's nastiest nuclear waste, as Ehmke admits. There would be jobs. Meanwhile, local officials are kept sweet with nuclear cash for their municipal budgets. A swimming pool bought the support of some. Opponents regard this as tainted money, and things can get nasty. Peter Ward, who has worked at Gorleben for twenty-five years, most recently as the chair of the works committee, says officials who support the project are met with death threats from protesters. "It's not what they expected when they became councilors of small rural villages."

But Gorleben's opponents do not all fit the stereotype of Green antinuke activists. I moved on from the spiky-haired protesters in the snow outside the waste store to a very different atmosphere—the warmth of a baronial hall, the ancestral home of the man who owns much of the land above the Gorleben salt dome. As we climbed the grand staircase of his mansion deep in the forest, Baron Andreas Graf von Bernstorff told me proudly that his forebears were Hanoverian Kings of England. In 1694, the man who twenty years later became King George I bought the hall, along with the surrounding villages. His portrait, along with those of a cavalcade of other leading Hanoverians, adorned the stairs.

In his upstairs sitting room, we settled in front of a blazing log fire with his wife and family. The avuncular but steely-eyed baron, dressed in a casual pullover, switched from contemplating his distinguished ancestry to his concern for his own legacy. He had an obligation, he said, to protect the family land and pass it on to future generations. So while his politics could not be more different from the Greens at his gate, he had found common cause with them against the nuclear threat.

It was 1977, he said, when West German nuclear officials first came striding down his long drive and knocked on his heavy wooden door. They told him they wanted a tenth of his fourteen-thousand-acre estate to build a nuclear reprocessing plant, and part of the salt dome beneath his land for a nuclear dump. They said the salt dome was the best place in the country to bury nuclear waste, and they offered thirty million deutsche marks. But Bernstorff was suspicious. Back then, his estate was right on the border with East Germany. All round the world, countries chose to build nuclear plants close to their borders. The choice looked more like geopolitics than geology. He was having none of it.

In 1720, King George had declared that his successors should keep the land "for future generations." Bernstorff said he still felt bound by that. "We were offered a lot of money, but we are responsible for the land. We can't shirk that responsibility," interjected his wife, Anna. Their neighbor sold land for the construction of the waste store, but "we refused to sell up. They didn't get a square meter from me," Bernstorff said. And he is not averse to direct action. He blocked the roads with tree trunks and gave his staff time off to protest whenever new flasks of waste arrived. "My father is a natural conservative," said his son, Fried, who will inherit the land. "But he joined Greenpeace and became a protester. We coordinate with the citizens' initiative, even though we have a different position. The church, which owns a lot of land, is with us too." He nodded to the village's Lutheran pastor, the bearded and bespectacled Eckhard Kruse, who was sitting quietly in an armchair near the fire.

Lawyers insist that, if it comes to it, the federal government could appropriate Bernstorff's land. "The baron can slow the process, but he can't stop it," I was told by Doerte Fouquet, a Berlin barrister who specializes in energy law. So the question becomes what the federal government will decide to do. A Final Storage Commission, made up of politicians and scientists, in 2016 advised on criteria for choosing a burial site.[3] It concluded that, despite the debacle at Asse, salt domes should remain one of the options, and it refused to rule out Gorleben. Its chairman, veteran parliamentarian Michael Muller, told me in Berlin a few weeks before publishing his advice, "We all believe deep geology is the best option, but I'm not sure if there is enough public trust to get the job done."

Many antinuclear groups boycotted his commission. They agree, in theory at least, that Germany must deal with its own waste, including reprocessing waste returned from France and the UK. Many I spoke to also agreed that deep burial was likely to be the best option. But the nub of the issue was trust. At root, the protesters did not trust even a government committed to shutting down the nuclear industry to handle the waste problem fairly. "We don't trust the independence of the process for deciding what to do with the waste," said Ehmke. He accused politicians of being unscientific in their choice of waste sites, but he distrusted the scientists too.

"I can understand why there is mistrust," said Flasbarth, the environment secretary, a little wearily. "It's because of the past. All I can say is we wouldn't go through such a long, intense, and costly process if we didn't want to select the safest site. We must not exclude Gorleben. That would be unfair on other potential sites." So Gorleben remains, as it has always been, the prime battleground in Germany's nuclear conflict—a conflict that shows little sign of abating.

Ehmke told me, "We cannot bury this waste here in northern Germany. There could be ten ice ages, with glaciers scraping away the rocks, before the waste is safe." But if not in northern Germany, where most of the waste has been produced, then where? He and many others were silent on that. "It appears that the long struggle over Gorleben has resulted in stalemate," said Andrew Blowers, a historian of nuclear power and its opponents, who has visited the place frequently. "It seems impossible to make progress with Gorleben in the mix, but also impossible without it."[4]

There is unlikely to be any turning back from the decision to banish nuclear power. But the politics of Germany's postnuclear future have yet to play out. If Germans ever thought that declaring an end to nuclear power would end their nuclear problems, they couldn't have been more wrong. The country is not alone among nuclear nations in its dilemmas over radioactive waste. But by so abruptly abandoning the nuclear project, it is doomed to pioneer solutions.

Chapter 25

Waste

Out of Harm's Way

More than seven decades after the dawn of the nuclear age, very little of the world's most dangerous radioactive waste has been successfully put out of harm's way, and the inventory keeps on growing. The European Union alone generates 1.4 million cubic feet annually, and the decommissioning of old power plants will add massively to this. Moreover, much that has been buried is now regarded as unsafe and may end up being disinterred. That includes the 100,000-plus drums buried in the Asse salt mine in Germany during the 1970s and perhaps also the 80,000 tons of waste containing almost a million curies of radioactivity (the equivalent of twenty Windscale fires) that the British dumped into the English Channel and North Atlantic, before the practice was banned in 1983.[1]

Nuclear waste is conventionally categorized as high-, intermediate-, or low-level waste. Low-level waste includes everything from discarded protective clothing to plant equipment and lab waste. This can be treated like any other hazardous waste, which in practice usually means burial in drums in landfills or concrete-lined trenches. Sites include, among many others, the Drigg dump near Sellafield in the UK and the trenches at Hanford.

Intermediate waste contains radioactive materials with half-lives long enough to require long-term incarceration. It includes many reactor components as well as chemical sludges and liquids from processing radioactive materials. The latter can often be solidified in concrete blocks. Intermediate waste can be buried safely in shallow graves, though anything containing plutonium will have to be disposed of deep underground because of its very long half-life. There is usually no reason to delay disposing of low and intermediate waste, yet more than 90 percent of it is awaiting a home.

The largest existing deep burial ground for intermediate waste is the US military's Waste Isolation Pilot Plant in salt beds in Carlsbad, New Mexico. WIPP could eventually take a quarter-million drums of waste. But it has had problems. A chemical explosion in 2014 sprayed waste in the tunnels. Some plutonium reached the surface through the ventilation system, and seventeen workers were contaminated. The military shut the tunnels for three years to clean up. The eventual cost of the accident, including keeping the dump open longer to catch up with the waste back-log, has been put at $2 billion.[2]

High-level waste is the nastiest stuff. It is either very radioactive and will stay so for a long time, or it generates heat and so requires keeping cool. Usually both. This category contains more than 95 percent of all the radioactivity in nuclear waste. Most of it is spent fuel—an estimated 200,000 tons currently around the world. Much of the rest is the highly radioactive waste liquids produced when reprocessing spent fuel. Some 170 million gallons are thought to be in store round the world, a third of them at Hanford.

High-level waste needs to be kept out of harm's way for thousands of years. It is generally agreed that the only safe way of doing that is to turn the liquids into solids and then bury it all deep underground, where water won't bring the radioactivity to the surface and future generations are un-likely to stumble upon it. There is disagreement, however, about whether this buried waste should be kept retrievable in case future technologies could make it safer sooner, or whether accessibility simply places a burden of guardianship on future generations.

Most high-level waste needs to be stored while it cools, often for up to a hundred years. Unfortunately, this has encouraged countries to put off even thinking about where it should go. As a result, according to the UN International Atomic Energy Agency (IAEA), none of the world's high-level waste currently has anywhere permanent to go.[3] The world is instead peppered with interim stores, some designed for the pur-pose, but others simply ad-hoc cooling ponds attached to power stations. America's ninety thousand tons or so of spent fuel, for instance, is mostly sitting in ponds at dozens of power stations or lockups. These include a store at Fort St. Vrain, north of Denver, Colorado, that has been sitting on the prairie like some outsize grain store since 1991, when the neigh-boring nuclear power plant shut and the governor of Idaho banned the shipment of its fourteen tons to another temporary store at the Idaho National Laboratory.[4]

This cannot go on forever. The federal government has long harbored plans to inter the nation's high-level waste inside Yucca Mountain, near the Nevada bomb testing grounds. In the 1990s, it dug a tunnel five hundred yards into the mountain but then went no further. The $100 billion Yucca plan was formally abandoned by the Obama administration in 2009, after geologists warned that a volcanic eruption could propel buried waste back to the surface. But President Trump has indicated he wants to revive the project. Maybe Yucca Mountain will make a comeback. But if not, with no alternatives on the horizon, then interim stores in places such as Fort St. Vrain could be in business not just for decades but for centuries.

In other countries, France has earmarked for its burial ground the clay beds at Bure, in a remote and thinly populated area in the northeast of the country. The main drawback may be the five-hundred-mile journey from its nuclear reprocessing and waste storage hub, which is at Cap La Hague, near the Channel ferry port of Cherbourg. The British government, as we saw in chapter 22, prefers the proximity of tunneling directly from Sellafield under the Lake District National Park. Russia is exploring a site in Krasnoyarsk, five thousand miles from Moscow on the Trans-Siberian Railway. Germany may or may not choose Gorleben. China won't pick a site until at least 2020. Japan can't find anywhere on its crowded and quake-prone islands able and willing to host a dump.

The countries that have gotten the furthest are Sweden and Finland. In early 2017, the Finns began constructing a "final disposal facility" for high-level waste, 1,600 feet down in granite rocks on Olkiluoto Island, in western Finland. It should open in 2023. Sweden has chosen a site at Forsmark, north of Stockholm.

Most nuclear nations think it is quite hard enough to find somewhere to bury their own radioactive waste. But some would like to sell their geological space to the world. Kazakhstan has at times fancied turning its uranium mines in the west of the country into an international high-level waste dump; alternatively it could repurpose the Semipalatinsk testing grounds.[5] Ukraine reckons that the lands around Chernobyl would provide a good cheap and neighbor-free burial environment, with plenty of domestic radioactive nasties close by to get the project started. South Australia pitched for the same market but found the local Aborigines hostile.

As the London-based World Nuclear Association summed up: "At the heart of their opposition were the memories of the British weapons tests at Maralinga, where many communities had been displaced and indeed many individuals had suffered directly from the side-effects of the blasts."[6]

||||

Much of the need for deep burial of nuclear waste arises from the plutonium content. The exceptionally long half-life of this metal requires keeping it out of harm's way for tens of thousands of years. But many atomic scientists think that plutonium should not be regarded as a waste at all. It is too valuable. They think spent fuel from reactors should be routinely reprocessed to extract the plutonium. That would also make the spent fuel easier to dispose of. Is that a good idea?

As we have seen, the first reactors—at Hanford, Ozersk, and Windscale —were constructed entirely for the purpose of making plutonium for atomic weapons. Reprocessing was routine. It was only when most reactors were built for power generation that the plutonium became an irritating by-product. Even then, reprocessing plutonium continued in some countries, with the idea of using it to fuel a new generation of power stations. Britain and France in particular continued with the reprocessing to fulfill this end, even when the need for more plutonium for weapons faded.

France made good headway. Roughly a sixth of all its electricity comes from burning recycled nuclear fuel, mostly in the form of mixed-oxide, or "mox," fuel. This is an amalgam of reprocessed plutonium and conventional uranium fuel, and has the advantage of being able to burn in a conventional PWR power station.

But British efforts proved expensive failures. An ambitious prototype plutonium-burning fast-breeder reactor that could generate more fuel than it consumed was abandoned thirty years ago after a damning safety audit. Then a mox manufacturing plant at Sellafield proved unworkable and was abandoned in 2011, with $2.6 billion spent.[7] The British were not inclined to press on when, with plenty of cheap uranium available, power station companies said they were not interested in buying any of the new, more expensive fuel that might be produced. The impasse is one reason for a growing plutonium stockpile round the world, which is arguably the worst of all worlds.

The Stockholm International Peace Research Institute estimates that the world is currently sitting on around 550 tons of plutonium. It is divided equally between civilian and military stores. Both the US and Russia have large military stocks, mainly inside existing nuclear weapons or extracted from decommissioned weapons. The US said in 2012 it had one hundred tons of plutonium. Tens of thousands of American plutonium "pits," mostly produced at Rocky Flats, are held at the PANTEX weapons assembly plant in Amarillo, Texas. They are stacked in underground bunkers, waiting to be either put into future bombs or disposed of. More US plutonium is stored at the Savannah River nuclear complex in South Carolina.

In 2000, Russia and America agreed to get rid of 37.5 tons of surplus military plutonium each, as part of a joint weapons reduction program. That would have removed from the global stockpile enough plutonium to make seventeen thousand nuclear weapons, more than the world's entire current weapons inventory. It would have made the world a safer place. The plan was to dispose of the plutonium by mixing it with uranium to make mox fuel for burning in power stations. The arrangement was dubbed "megatons to megawatts," a reworking of the old theme of "atoms for peace."

The deal was "reconfirmed" in 2010. But while Russia pushed ahead with a mox plant, the United States changed tack. President Obama was not keen on nuclear power, so had little enthusiasm for making mox fuel. In 2014, with costs of the US mox plant spiraling beyond $8 billion, he suspended construction. He proposed an alternative plan, to blend the weapons plutonium instead with lower-grade civilian stocks of plutonium, and then bury it all at the Waste Isolation Pilot Plant in New Mexico.[8] But Russian president Vladimir Putin cried foul. He said the mixing could be reversed, so that the weapons material could one day be reclaimed. That, he said, wasn't the deal. Putin announced he would not, after all, be burning any of his weapons plutonium until a new agreement was reached.[9]

In 2017, it was not clear if the initially chummy relations between Putin and newly elected President Trump would restart the deal, or whether Trump would want to hold on to his plutonium for a promised resumption of nuclear weapons production. But Putin had raised a valid question. As we saw in chapter 6, at the plutonium mountain, buried plutonium can be recovered. Waste dumps can, in theory at least, become future plutonium "mines" for people up to no good.

||||

With no deal between the superpowers on getting rid of military plutonium, the world's stockpile of one of the world's nastiest substances keeps
growing. And not just within the confines of military stores. One of the
things that alarmed me most during my investigation of the legacy of the
atomic age was the discovery that the world's largest store of plutonium
is not in some military bunker guarded by the firepower of a superpower,
but in a warehouse at Sellafield on the shores of the Irish Sea, guarded by
a police force, Britain's Civil Nuclear Constabulary. Here, at last count,
sits about 130 tons of powdered plutonium dioxide, sealed in thousands
of stainless-steel cans, each about the size of a lunch box. It is enough to
make twenty thousand Nagasaki-size bombs.

This gigantic stockpile is the product of the continued reprocessing of
spent fuel at Sellafield; it continues to grow by about four tons a year. By
2020, when the last reprocessing is scheduled to be completed, it should
exceed 150 tons.[10] The intention has always been to turn it into "mox" or
some other type of advanced nuclear fuel. Enthusiasts at the heart of government have pushed this agenda. The government's former chief scientist
for energy, David MacKay, who died in April 2016, once argued that it
could run the country's electricity grid for five hundred years.[11] But with
no sign of any investment to turn that dream into reality, the stockpile just
sits there. This seems like madness to me.

Plutonium dioxide powder is not the purest form of plutonium. But
the Royal Society, Britain's most venerated science institution, warned
in 2007 that the store was nonetheless a "major security risk," vulnerable to terrorist attack. Terrorists might break in and steal some of the
powder, which could quite easily be made into a crude nuclear device.
Or they might simply launch a weapon into the building, which would
blast the deadly powder into the air across northern England. The author
of that report, Geoffrey Boulton, of Edinburgh University, says that the
risk remains very real. The problems he identified had not been fixed, he
told me when I met him a decade later in an upstairs room at the Royal
Society's palatial offices in central London.[12]

Boulton is a geologist not given to hyperbole. But he was perplexed
and angry that successive governments had largely ignored his report's recommendations. He and his panel of five professors had set out an urgent

program for making the stockpile safer. He had thought ministers would take the warning seriously. Yet "the government has only delivered on one of our four immediate priorities," he told me. Meanwhile, the presence in Europe of ever more terrorists made the risks of something bad happening to the stockpile ever greater.

The one advance made by the government since Boulton's report has been to complete a new building to store the plutonium. Opened in 2010, it is three hundred feet long and substantially larger than the Albert Hall. Its managers insist it is secure, but they are reluctant even to identify the location of the building within the Sellafield complex. No doubt that is part of their security strategy. I won't undermine it by identifying where the building is. But if I can find out, then others can. And the evidence of the past is that security may not be as good as they hope.

In 1995, some two hundred Greenpeace supporters invaded Sellafield and leapt over the fence around the plutonium store, spraying the word "bollocks" on the wall. British Nuclear Fuels Ltd. (BNFL), which ran the plant at the time, claimed "there was no danger to the facilities." But its company secretary until the previous year, Harold Bolter, later called that claim "asinine." How could there have been no danger, he said, when "a couple of hundred demonstrators, who might have been armed terrorists for all BNFL knew," had poured onto the site unimpeded and reached the plutonium store? Or when, as a Royal Commission report on the British nuclear industry had argued as long ago as 1976, "plutonium appears to offer unique and terrifying potential for threat and blackmail against society."[13]

Boulton told me he hoped the store's managers had learned from the past and considered the risks posed by today's generation of terrorists and today's terrorist technology. "I assume that thought has been given to what a malicious individual might do to the store's physical integrity —with a shoulder-mounted missile, or wearing a suicide vest without thought of their own survival," he said. A successful small-missile strike, he warned, "could project plutonium powder into the atmosphere, with the potential for widespread dispersal." It would be a terrorist Chernobyl. Did he know if the nuclear authorities had considered the matter seriously, I asked. Maybe they had, he said, but if they had found an answer to the threat, they hadn't told him.

Boulton's 2007 report set out four "immediate priorities" for defusing Britain's plutonium time bomb. The first, upgrading security, may or may not have been achieved at the new store. But he also called for a halt to further growth of the stockpile. Instead, the stockpile has grown by a further forty-five tons. And he called for immediate work on turning the powder into mox fuel. He argued that even if mox fuel was never burned to make energy, the mixing would keep plutonium in a safer form than it is at present. But the planned mox fuel-making plant has been canceled.

In truth, there has been no British conspiracy here. It is just an extraordinary, embarrassing, dangerous, and expensive fiasco, arising from Britain's thwarted ambition to turn spent fuel into a big new moneymaking industry. The stockpile has kept on growing because policymakers have been on autopilot. Nobody has called a halt to reprocessing to make more plutonium because nobody has wanted to admit that the world's biggest stockpile of perhaps the world's most dangerous element is useless.

Nothing that had happened since the publication of his report had changed Boulton's view that "the status quo of continuing to stockpile a very dangerous material is not an acceptable long-term option."[14] I have to agree. Having spent several hours wandering around the perimeter of Sellafield, I don't believe it would be too difficult for a terrorist either to identify where the plutonium is stored, or to aim a rocket at the stuff.

Conclusion

Making Peace in Nagasaki

What is the future of nuclear power? Does it have a future? As I write in late 2017, the commercial outlook for the industry is bleak. Looking back, it's clear that nuclear energy has prospered only with heavy government financial support. Whenever market forces and private investment have gotten involved, it has usually failed. British Nuclear Fuels went bankrupt. French power companies would have gone the same way but for the support of the government. Investment in US nuclear power evaporated after Three Mile Island. In early 2017, Westinghouse, the Pittsburgh pioneer of the world's workhorse pressurized-water reactors, filed for bankruptcy after the failure of its latest reactor design. Its Japanese parent company, Toshiba, was also in dire financial straits. Six months later, utilities pulled the plug on two of the four power stations then under construction in the United States.

Some advocates continue to argue that boom times could be just around the corner, that the drive to phase out fossil fuels should herald the dawn of a new nuclear age. Even some environmentalists have argued that nuclear power, as a proven, large-scale source of low-carbon energy, might be a good partner for renewables—at the very least, a valuable backup when the winds falter and the sun goes down.[1] Sadly, as Germany's federal environment secretary, Jochen Flasbarth, explained to me when we met in Berlin, that won't work. Nuclear reactors cannot easily be turned on and off. Nuclear must provide "baseload" power, or nothing, he said. "It is too inflexible to work well with renewables."

Engineers still talk of radical new designs—modular reactors might be the next big thing, or reactors that could run on thorium or sodium. Perhaps the fast breeder could make a comeback and eat up all that plutonium. Billions are still being spent on researching fusion reactors. But the date when they might deliver serious energy keeps receding into the future. Nuclear power increasingly looks like a twentieth-century solution to a twenty-first-century problem. The world seems to be going in the opposite direction.

While renewables get cheaper, nuclear energy gets ever more expensive. When Britain decided to build its first nuclear power station in more than thirty years, at Hinkley Point, the government ended up signing a contract with EDF, the French company, guaranteeing to buy the power at a price far above the cost of other sources. Only months later, as I write, the construction bill has risen another $2.8 billion to more than $26 billion.

And the potential liabilities from a major nuclear disaster always loom in the background. The Three Mile Island cleanup bill of a billion dollars was a cinch. The Fukushima cleanup at the time of writing is estimated at $180 billion and may not have stopped rising. TEPCO, the giant corporation that ran the plant, survives only because the Japanese government nationalized it. The Chernobyl cleanup will take another century at least. Even if it is not your accident, a disaster anywhere in the world can trigger an orgy of extremely expensive regulatory rethink. For private investors such risks are too much to bear. The nuclear industry is uninsurable. Governments have to underwrite everything.

But whatever the finances, a major industry like this can prosper only with public consent—and public opposition to all things nuclear continues to grow. Germany has bailed out after long and bitter divisions. Six years on from Fukushima, local opposition meant that only five reactors at three of Japan's fifty-plus nuclear power plants had resumed operations. The new South Korean president, Moon Jae-in, announced a phase-out of the country's previously booming nuclear power industry, citing fears of a future Fukushima-style disaster.[2] Even France, the Western country most enthusiastic about nuclear energy, is getting cold feet. New president Emmanuel Macron was elected on a commitment to stick with his predecessor's policy of reducing nuclear power's share of the country's energy supply from 75 percent to 50 percent. That leaves as significant players in the civil nuclear game only state-owned enterprises in Russia, which has been increasingly keen to sell its technology around the world, and China, whose thirty-seventh civil reactor went on line at the end of 2017.

On many safety issues I have some sympathy with the industry. Too much of the invective of antinuclear activists bears little relationship to truth. Its scientific probity is sometimes on a par with the worst excesses of the climate skeptics. On nuclear issues, there has long been a post-truth world out there. When I headed for Mayak or Chernobyl, Fukushima or Sellafield, Rocky Flats or Hanford, friends and editors asked me whether I wasn't afraid of the radiation. No, I wasn't. On my return from the Mayak exclusion zone, I liked to explain that I had to wear a protective suit, but against ticks, not radiation. My questioners seemed bored by that. They much preferred the news that if the scanners at the gates when I was leaving Chernobyl had found me to be radioactive, I would not have been allowed to leave the exclusion zone. I could have joined the radioactive wolves in the forest.

Over my years as an environmental journalist, I have written my share of scare stories about radiation. But on my journeys for this book I was prepared to trust the scientists and doctors I was meeting and to believe the stats. But of course, I had no personal way of knowing if I was right to be sanguine. Without a Geiger counter, my senses had no way of telling me of any threat. I can smell smog and intuitively understand something of its danger. But radiation is different. We don't have a radiation ear.

That is the crux for public consent to nuclear power. If we can't trust our own senses, then it is vital that we can trust the experts. The trouble is that over half a century and more, the nuclear industry has gratuitously and comprehensively lost that trust. Nuclear plant operators, their owners, governments, and regulators have too often covered up, tried to hide the scale of their failures, obfuscated about accidents, gone AWOL in an emergency, and gone into denial about the cost of cleanup. The evidence of Sellafield's former company secretary, Harold Bolter, is that they don't even tell the truth to their colleagues and bosses. That means they fail to learn from their mistakes and fail to reassure the rest of us.

Many supporters of nuclear energy charge that the world is afflicted by "radiophobia." They say we must lose our irrational fears and embrace nuclear power. But not so fast. Irrationality about nuclear matters afflicts both sides of the debate. Our nuclear masters' legacy of military secrecy and current bunker mentality skew and undermine how they manage radioactive materials and processes. It can sometimes push regulators toward policies that vastly increase the cost of nuclear activities by "overdesigning" everything. But it can also make engineers—and some regulators—

cavalier, arrogant, secretive, cynical, complacent, and dismissive of concerns. Many of the accidents I have described in this book show this in spades.

The word radiophobia is also used to suggest that the industry has no responsibility for society's fears. It excuses their failings and makes the nuke-fearers into hypochondriacs and irrationalists. But the industry cannot be allowed to absolve itself of responsibility for any fallout from its activities that is not radiological. Through mishandling its power plant at Fukushima and failing to explain to the world what was going on, TEPCO was responsible for the suicides and depressions and indeed radiophobia among the tens of thousands of people evacuated after the accident. I agree that those who fan the flames of fear by telling lies deserve to be called out. But our fears are real. They too are part of the fallout.

The story of the nuclear age has been a tragedy: a saga of arrogance and hubris, of misplaced and betrayed trust. The result is often a paranoid and hyperbolic discourse about all things nuclear. This is the Anthropocene on steroids, a world where technology has conferred power but also created fear and a creeping policy paralysis. Can we ever talk sense about nukes? I begin to wonder. If not, then I think the technology becomes unsustainable in a democracy.

So, at the end of my journey through the atomic age, I have reached what may seem contradictory conclusions. First, that while much that went on in the Cold War was dangerous and duplicitous, most civilian nuclear activities are safe—or at least far less dangerous than is often supposed. Fukushima is the emblem of that: the deaths were all to do with fear and flight rather than radiation. But second, it seems to me that societies have a perfect right to turn their back on nuclear technologies if the experts fail to win their trust. It may be true, as Franklin Roosevelt put it, that we often have nothing to fear but fear itself. But the fear is real, nonetheless. And, if after more than half a century, nuclear protagonists have failed to still that fear, then probably they never will.

Maybe we have to concede that this is a dying industry. The atomic age looks like it is over. The future for nuclear energy may be simply for us to see out the lives of existing plants and deal with their environmental legacies as best we can. Oh, and to get rid of nuclear weapons.

||||

To end this journey, I finished as I began, in Japan—the only place where the world has experienced something of the true horror of nuclear conflagration. I went to Nagasaki. The city is often the forgotten site of atomic destruction. Hiroshima got most of the headlines. But at least seventy thousand people died when the Americans' plutonium device exploded 1,600 feet above the city, a height that had been chosen to maximize the impact of the blast and firestorms that burned all the city's wooden buildings. It was this second bomb that finally persuaded the Japanese emperor to countermand his generals and surrender.[3]

Seventy-one years later, a courteous and bespectacled Tatsujiro Suzuki met me in his small office on the campus of Nagasaki University, less than a mile from ground zero that day. In 2014, he had been appointed the director of the Research Center for Nuclear Weapons Abolition, an organization inspired by President Obama's call in 2009 for a world without nuclear weapons.[4] The center was, Suzuki said proudly, the first academic institute on peace in the city. But his arrival there was part of the fallout from Fukushima.

At the time of the Fukushima accident, Suzuki was vice chairman of the Atomic Energy Commission, working for the Japanese government. But the accident shocked him. "After Fukushima, my attitude to nuclear energy changed," he told me. "As a member of the commission, I felt I was partially responsible for what happened. I couldn't carry on being a pronuclear advocate." He had not become antinuclear, he said. "But severe accidents like Fukushima make it very difficult to persuade the public that the next plant will be safe. We said that last time." For him, perhaps most important of all, "the connection with nuclear weapons is still the Achilles' heel of civil nuclear power." As long as countries use the technology for generating energy, "you can never eliminate the risk of nuclear proliferation."

In his new academic role, Suzuki was pushing for global nuclear disarmament. He wanted to begin by helping to create a nuclear-free zone in northeast Asia. He feared bomb construction in North Korea would encourage Japan to develop its own nuclear weapons. A few weeks later in the US, President-elect Donald Trump appeared to encourage that when he called on Japan to take a bigger role in protecting itself. Suzuki feared that a new generation of leaders in Tokyo would forget the terror of nuclear war and be swayed by such talk. Outside Nagasaki and Hiroshima, he said, the average Japanese person's knowledge of nuclear weapons is very poor. "The media no longer report the annual commemorations. We need to transfer the message about the horrors of the bomb to the next generation."

Suzuki's institute had set up a "citizens' database" on fissile material that might be used in bombs. His highest priority was the stockpiles of plutonium that he felt were the most tangible risk of proliferation. Managing plutonium had been his academic specialty at Tokyo University years before. "I once believed, like most people in the nuclear industry, that reprocessing spent fuel to get plutonium was essential to the future of nuclear power," he told me. "But I don't think that now."

I asked what he thought about the world's largest stockpile, in a shed next to the Sellafield reprocessing plant in the UK. Separating pure plutonium in reprocessing plants created a risk of proliferation among nations, he said. It also increased the risk that a terrorist would create a "dirty bomb" by blowing up a plutonium stockpile or stealing some and exploding it in the heart of a city. "I think plutonium reprocessing should end in all states," he said. "We don't need it." All the world's existing plutonium should be put under strict international oversight to prevent its misuse. That should include the eleven tons in store in Japan, and the forty-two tons due to be returned to Japan from reprocessing in France and Britain, as well as the big stockpiles in Britain, the US, and Russia.

Suzuki backed Nagasaki's peace-minded mayor, Tomihisa Taue, who wanted to invite North Korean officials to a meeting of Mayors for Peace, an international network that includes more than seven thousand cities. "We are calling for [nuclear weapons] abolition as a message from the heart, but we need logic too. We need to find ways to reduce our reliance on nuclear deterrence," Suzuki said. Where better to start, he said, than one of the two places on the planet hit by such a "deterrent."

Nagasaki University's campus was rebuilt from the rubble of the old Nagasaki Medical College in the suburb of Urakami, which took the full force of the American bomb. Reminders of the horrors were all around us. The college lost nine hundred students and staff. Suzuki showed me the small campus museum that commemorates his academic predecessors who raced to rescue their compatriots that day.[5]

One of the ironies of Nagasaki's fate was that, of all Japanese cities, it had long been among the most open to the outside world. Nearby was the rebuilt Urakami Cathedral, a bastion for Christians in Japan since Jesuits set up in Nagasaki in the 1580s. Only the toppled bell tower remained from 1945. On the site of the city's old prison there was now a peace park. During my visit, they were preparing for the annual commemoration, setting up large tents, checking the sound system, and erecting displays showing in unsparing detail the carnage of that day seventy-one years before. At the commemoration, Mayor Taue called for the international community to use its "collective wisdom" to rid the world of nuclear weapons.[6]

I looked for ground zero, the area of land directly below where around thirteen pounds of plutonium had exploded. Much of it had been excavated for the renovation of a stream. A few of the stones close by still bore the scars from the heat flash. But hung up along the stream's banks, there was a line of forty-five brightly colored murals produced by local children—some depicting the moment of destruction, but most showing rainbows of hope and handshakes of peace.

I complete this book with ferocious rhetoric between the United States and North Korea echoing in my ears from news broadcasts almost every day. It is profoundly worrying. In the West, North Korea's leader, Kim Jong-un, is characterized as the villain in his desire, as he sees it, to secure his country's future by acquiring atomic weapons. What we seem to forget is the terms of the Nuclear Non-Proliferation Treaty signed by all nuclear nations and most others, which came into force in 1970. It was essentially a deal under which non-nuclear nations agreed to forgo joining the nuclear club in return for a promise from the nuclear nations to pursue nuclear disarmament. The ultimate aim was to return to a world without such weapons.

Some progress was made. The tragedy is that almost half a century on, disarmament has ground to a halt. If countries such as North Korea now say that the deal has been broken and claim the right to develop their own weapons, then nuclear nations now have only themselves to blame.

In October 2016, in an effort to revive the drive to disarmament, the General Assembly agreed "to convene a UN conference to negotiate a legally binding instrument to prohibit nuclear weapons, leading towards their total elimination." In June 2017, more than 120 nations signed the ban. But it was the abstainers that drew more attention. They included all nine of the nations with the planet's current arsenal of fifteen thousand nuclear weapons: the US, Russia, China, France, Israel, the UK, India, Pakistan, and North Korea.

The Obama administration had lobbied strongly against the ban. "How can a state that relies on nuclear weapons for its security possibly join a negotiation meant to stigmatize and eliminate them?" asked its UN ambassador, Robert Wood.[7] Maybe he had forgotten that, during a visit to Hiroshima the previous May, his boss, President Obama, had called for a "moral revolution" against nuclear weapons. "We must have the courage to escape the logic of fear and pursue a world without them," Obama had said.[8] Memories are short.

I have kept technical terms to a minimum and explain them when I first use them. Even so, a list of the ones that are unavoidable, and what they mean, could come in handy:

Chain reaction · a nuclear reaction in which neutrons bombard atoms such as uranium, ejecting more neutrons that repeat the process

Criticality · a situation in which a chain reaction is set off

Decay · when a radioactive isotope turns into another isotope, giving off radiation

Decommissioning · taking apart a power station or other structure

Dose · radiation absorbed by living tissue, measured here in millisieverts

Epidemiologist · scientist measuring disease rates

Fast breeder · nuclear reactor that produces more fuel than it consumes

Fission · splitting atoms to release energy; for instance, in an atomic bomb or nuclear reactor

Fusion · combining atoms to release energy; for instance, in a hydrogen bomb

Geiger counter · instrument for measuring radiation

Half-life · time it takes for half of newly generated isotopes to decay

Isotope · one of a number of types of atoms of the same element that contain different numbers of neutrons

Liquidator · worker sent to scene of nuclear disaster to make plant safe for cleanup

Plutonium economy · imagined future world reliant on plutonium for energy

Radiation · the energy from radioactive emissions

Radioactive waste · any waste containing radioactive materials. Usually categorized as low-, intermediate-, or high-level, depending on the intensity of the radiation emitted and its half-life

Radioactivity · emissions of radiation during decay of an isotope,
    measured here in curies

Reactor · container in which fission is orchestrated to generate energy
    and fission products such as plutonium

Reprocessing · chemical processes to retrieve fission products from reactor
    spent fuel

Spent fuel · nuclear fuel that has passed through a reactor

Wigner energy · displacement of atoms in graphite inside a reactor.
    Potentially dangerous

Acknowledgments

Like most journalists, I try to mention and attribute my main sources in the text. I won't repeat all those names here. But a few deserve special mention for their support on my journeys.

Thanks to Malgorzata Sneve and others at the Norwegian Radiation Protection Authority for help in understanding Soviet nuclear legacies and making key Russian introductions for me. In Chelyabinsk, Mayak's Yuri Mokrov was a beacon of candid help. In the Chernobyl exclusion zone in Ukraine, Gennady Laptev and Sergey Gaschak were invaluable, along with Nick Beresford, who invited me to his wildlife workshop. The vodka was courtesy of Markeyevych Federovych. For digging into the Semipalatinsk archives and translating the findings, I thank Kazbek Apsalikov and Sholpan Zhakupova at the Institute of Radiation Medicine and Ecology. At the Kurchatov Institute, thanks to Sergey Lukashenko.

At Rocky Flats, LeRoy Moore, Jon Lipsky, and Harvey Nichols were hugely helpful, along with Tiffany Hansen of the Downwinders and, on the federal side, Scott Surovchak and David Lucas. Elsewhere in Colorado I greatly enjoyed my day on the silo trail with Bill Sulzman. In Seattle, Tom Carpenter was most valuable on Hanford. In the hinterland of Sellafield, my guide was Martin Forwood of Cumbrians Opposed to a Radioactive Environment. Thanks also to Andrew Pearson for my tour within the fences of the complex.

Yoshimi Miyake and Murakami Akira at Akita University introduced me to Baba Isao. Thanks also for a second Fukushima journey with American students organized by William McMichael, as well as to Keith Franklin at the British Embassy in Tokyo and Ken Nollet at Fukushima Medical University. In Nagasaki, I spent valuable time with Shunichi Yamashita and Tatsu Suzuki at Nagasaki University. In Germany, I gratefully acknowledge the assistance of Clean Energy Wire in organizing a press tour. Peter Ward offered valuable post-trip perspectives.

Many experts and campaigners have helped enlighten me about nuclear and medical matters, though of course any errors are my own. They include that most forensic of campaigners, Walt Patterson, plus Gerry Thomas at Imperial College London, Andrew Blowers at the Open University, and Oxford's Wade Allison, as well as Stephen Tyndale, Mark Lynas, Ted Nordhaus, Michael Shellenberger, John Large, and Geoffrey Boulton at the Royal Society.

Many of my colleagues over the years at *New Scientist* have filed stories and unlocked contact lists that I have drawn on. Among them have been Roger Milne, Tom Wilkie, Catherine Caufield, Rob Edwards, Michael Kenward, and the late Ian Anderson. Some material here has appeared in different forms elsewhere, thanks to commissions by Roger Cohn at *Yale Environment 360*; Mico Tatalovic, Catherine Brahic, and Sally Adee at *New Scientist*; Greg Neale at *Resurgence*; and Sigrid Rausing at *Granta*, whose request for an article on Sellafield was the inspiration for the wider study. Those commissions helped pay for my journeys.

Thanks also, of course, to my book editors in Boston and London, Amy Caldwell, Laura Barber, and Ka Bradley, for their enthusiasm, insights, and perseverance.

Notes

Abbreviations used below
CTBTO—Comprehensive Nuclear-Test-Ban Treaty Organization
IAEA—International Atomic Energy Agency
ICRP—International Commission on Radiological Protection
NRDC—Natural Resources Defense Council (US)
UNEP—United Nations Environment Programme
UNSCEAR—United Nations Scientific Committee on the Effects
    of Atomic Radiation
WHO—World Health Organization
WNN—World Nuclear News

Introduction   Anthropocene Journey
1. William Standring et al., "Overview of Dose Assessment Developments and the Health of Riverside Residents Close to the Mayak PA Facilities, Russia," *International Journal of Environmental Research and Public Health* (2009), doi: 10.3390/ijerph6010174.
2. Mira Kosenko, "Where Radiobiology Began in Russia: A Physician's Perspective," Defense Threat Reduction Agency, 2010, https://pdfs.semanticscholar.org /8ed5/8f2548e0cf3c58de7c24cbfd55291d6faf1a.pdf.
3. Brian Jay, *Britain's Atomic Factories: The Story of Atomic Energy Production in Britain* (London: Her Majesty's Stationery Office, 1954).
4. Robert Hunter, *The Greenpeace Chronicle* (London: Picador, 1980).
5. "Media Note: Anthropocene Working Group," University of Leicester, UK, http://www2.le.ac.uk/offices/press/press-releases/2016/august/media-note -anthropocene-working-group-awg.
6. Colin Waters et al., "Can Nuclear Weapons Fallout Mark the Beginning of the Anthropocene Epoch?," *Bulletin of the Atomic Scientists* 71, no. 3 (2015): 46–57.

Chapter 1   Hiroshima: An Invisible Scar
1. John Hersey, *Hiroshima* (London: Penguin Books, 1946).
2. Ibid.
3. Yehoshua Socol et al., "Atomic Bomb Survivors Life-Span Study," *Dose-Response* 13, no.1 (2015), doi: 10.2203/dose-response.14–034.Socol.
4. Dale Preston et al., "Effect of Recent Changes in Atomic Bomb Survivor Dosimetry on Cancer Mortality Risk Estimates," *Radiation Research* 162, no. 4 (2004): 377–89.
5. "'Rain of Ruin' Threat to Japan," *Manchester Guardian*, August 7, 1945, https://www.theguardian.com/theguardian/1945/aug/07/fromthearchive1.
6. Hersey, *Hiroshima*.

Chapter 2   Critical Mass: MAUD in the Nuclear Garden

1. Robert Jungk, *Brighter Than a Thousand Suns* (London: Penguin Special, 1960).
2. Ibid.
3. Fred Roberts, *60 Years of Nuclear History* (Charlbury, UK: Jon Carpenter, 1999).
4. David Cohen, "Secret Fission Papers Were Too Hot to Handle," *New Scientist* (June 6, 2007), https://www.newscientist.com/article/mg19426073-900-secret -fission-papers-were-too-hot-to-handle/.
5. Jungk, *Brighter Than a Thousand Suns*.
6. Susan Williams, *Spies in the Congo: The Race for the Ore That Built the Atomic Bomb* (London: Hurst, 2016).
7. Leona Marshall Libby, *The Uranium People* (New York: Crane, Russak, 1979), cited in *American Scholar*, Februay 29, 2016, https://theamericanscholar.org/cpb -spring-2016/#.
8. Mike Rossiter, *The Spy Who Changed the World: Klaus Fuchs and the Secrets of the Nuclear Bomb* (London: Headline, 2014).
9. Jungk, *Brighter Than a Thousand Suns*.
10. Rossiter, *The Spy Who Changed the World*.

Chapter 3   Las Vegas: Every Mushroom Cloud Has a Silver Lining

1. Matt Blitz, "Miss Atomic Bomb and the Nuclear Glitz of 1950s Las Vegas," *Popular Mechanics*, April 26, 2016, http://www.popularmechanics.com/science /energy/a20536/who-are-you-miss-atomic-bomb/.
2. Nathan Hodge and Sharon Weinberger, *A Nuclear Family Vacation: Travels in the World of Atomic Weaponry* (London: Bloomsbury, 2008).
3. "Interview with Robert Friedrichs," Nevada Test Site Oral History Project, University of Nevada, 2005, http://digital.library.unlv.edu/api/1/objects/nts/1226 /bitstream.
4. Masako Nakamura, "'Miss Atom Bomb' Contests in Nagasaki and Nevada: The Politics of Beauty, Memory, and the Cold War," *U.S.-Japan Women's Journal* 37 (2009): 117–43.
5. Pravin Parekh et al., "Radioactivity in Trinitite Six Decades Later," *Journal of Environmental Radioactivity* 85 (2006): 103–20.
6. Andrew Blowers, *The Legacy of Nuclear Power* (Abingdon, UK: Routledge, 2017).
7. Hodge and Weinberger, *A Nuclear Family Vacation*.
8. "Atomic Tests in the Nevada Test Site Region," US Atomic Energy Commission, 1955, https://www.fourmilab.ch/etexts/www/atomic_tests_nevada/.
9. Lee Torrey, "Disease Legacy from Nevada Atomic Tests," *New Scientist* (November 1, 1979): 336.
10. Kevin Watanabe, "Cancer Mortality Risk Among Military Participants of a 1958 Atmospheric Nuclear Weapons Test," *American Journal of Public Health* 85 (1995): 523–27.
11. Carl Johnson, "Cancer Incidence in an Area of Radioactive Fallout Downwind from the Nevada Test Site," *Journal of the American Medical Association* 251 (1984): 230–36.
12. "Dirty Harry," CTBTO Preparatory Commission, https://www.ctbto.org /specials/testing-times/19-may-1953-dirty-harry/.
13. Howard Ball, "Downwind from the Bomb," *New York Times Magazine*,

February 9, 1986, http://www.nytimes.com/1986/02/09/magazine/downwind -from-the-bomb.html?pagewanted=all.

14. Joseph Bauman, "Atomic Shame," *Deseret News*, October 28, 1990, http://www.deseretnews.com/article/129361/ATOMIC-SHAME.html.

15. "Negligence Ruling on U.S. Atom Tests Overturned," Associated Press, April 22, 1987, http://www.nytimes.com/1987/04/22/us/negligence-ruling-on -us-atom-tests-overturned.html.

16. Bob Harris, "The Conqueror and Other Bombs," *Mother Jones*, June 9, 1998, http://www.motherjones.com/politics/1998/06/conqueror-and-other-bombs.

17. "The United States Nuclear Testing Programme," CTBTO Preparatory Commission, https://www.ctbto.org/nuclear-testing/the-effects-of-nuclear-testing /the-united-states-nuclear-testing-programme/.

18. "The US Air Force's Commuter Drone Warriors," *BBC News*, January 8, 2017, http://www.bbc.co.uk/news/magazine-38506932.

Chapter 4   Pacific Tests: *Godzilla* and the *Lucky Dragon*

1. Roberts, *60 Years of Nuclear History*; and Catherine Caufield, *Multiple Exposures: Chronicles of the Radiation Age* (London: Secker & Warburg, 1989).

2. Wade Allison, *Radiation and Reason: The Impact of Science on a Culture of Fear* (Oxford, UK: Wade Allison, 2009).

3. Robert Newman, *Enola Gay and the Court of History* (New York: Peter Lang, 2004).

4. Steven Simon et al., "Fallout from Nuclear Weapons Tests and Cancer Risks," *American Scientist* 94 (2006): 48–57, https://www.cancer.gov/about-cancer/causes -prevention/risk/radiation/Fallout-PDF.

5. Nevil Shute, *On the Beach* (London: Book Club, 1957).

6. Hodge and Weinberger, *A Nuclear Family Vacation*.

7. "Commodore Ben H. Wyatt Addressing the Bikini Island Natives," National Museum of American History, http://americanhistory.si.edu/collections/search /object/nmah_1303438.

8. Ruth Levy Guyer, "Radioactivity and Rights," *American Journal of Public Health* 91, no. 9 (September 2001): 1371–76, https://www.ncbi.nlm.nih.gov/pmc/articles /PMC1446783/.

9. Caufield, *Multiple Exposures*.

10. Guyer, "Radioactivity and Rights."

11. Autumn Bordner et al., "Measurement of Background Gamma Radiation in the Northern Marshall Islands," *PNAS* 113 (2016), doi: 10.1073/pnas.1605535113.

12. Malgorzata Sneve and Per Strand, *Regulatory Supervision of Legacy Sites from Recognition to Resolution: Report of an International Workshop* (Oslo: Norwegian Radiation Protection Authority, 2016).

13. Woodford McCool, "Return of the Rongelapese to Their Home Island," Atomic Energy Commission, 1957, https://web.archive.org/web/20070925185914 /http://worf.eh.doe.gov/ihp/chron/A43.PDF.

14. "Evacuation of the Rongelap," Greenpeace International, http://www .greenpeace.org/international/en/about/history/mejato/; and Hodge and Weinberger, *A Nuclear Family Vacation*.

15. David Kattenburg, "Nuclear Paradise Lost," *Green Planet Monitor*,

November 16, 2013, https://www.greenplanetmonitor.net/conflict-and-environment/stranded-on-bikini/.

16. Coleen Jose et al., "This Dome in the Pacific Houses Tons of Radioactive Waste—and It's Leaking," *Guardian*, July 3, 2015, https://www.theguardian.com/world/2015/jul/03/runit-dome-pacific-radioactive-waste.

17. Sneve and Strand, *Regulatory Supervision of Legacy Sites.*

18. Rob Taylor, "Coral Flourishing at Bikini Atoll Atomic Test Site," Reuters, April 15, 2008, http://uk.reuters.com/article/us-bikini-idUKSYD29057620080415.

19. Keith Moore, "Nuclear Test Veteran Who Flew Through a Mushroom Cloud," *BBC History*, November 8, 2012, available at http://www.bbc.co.uk/history/0/20105140.

20. Rob Edwards, "Written Out of History," *New Scientist* (May 18, 1996), https://www.newscientist.com/article/mg15020302-200/.

21. "8 November 1957—Grapple X," CTBTO, https://www.ctbto.org/specials/testing-times/8-november-1957-grapple-x.

22. Rob Edwards, "300 Islanders Accuse UK Government of Exposing Them to A-Bomb Fallout," *Sunday Herald*, October 22, 2006, http://www.robedwards.com/2006/10/300_islanders_a.html.

23. Al Rowland et al., *New Zealand Nuclear Test Veterans' Study—a Cytogenetic Analysis*, report to New Zealand Nuclear Test Veterans' Association, 2007, http://www.massey.ac.nz/~wwpubafs/2007/Press_Releases/nuclear-test-vets-report.pdf.

24. "British Atomic Testing," ABC Radio transcript, June 2, 2001, http://www.lchr.org/a/36/9m/maralinga2.html.

25. Rob Edwards, "Britain Indicted for Cold War Crimes," *New Scientist* (February 8, 1997), https://www.newscientist.com/article/mg15320680-200-britain-indicted-for-cold-war-crimes/.

26. "French Nuclear Tests: 30 Years of Lies," *Nuclear Monitor*, 1998, https://www.wiseinternational.org/nuclear-monitor/487/french-nuclear-tests-30-years-lies.

27. "France's Nuclear Testing Programme," CTBTO, https://www.ctbto.org/nuclear-testing/the-effects-of-nuclear-testing/frances-nuclear-testing-programme.

28. "The Soviet Union's Nuclear Testing Programme," CTBTO, https://www.ctbto.org/nuclear-testing/the-effects-of-nuclear-testing/the-soviet-unionsnuclear-testing-programme.

Chapter 5   Semipalatinsk: Secrets of the Steppe

1. Roman Vakulchuk and Kristiane Gjerde, *Semipalatinsk Nuclear Testing: The Humanitarian Consequences* (Oslo: Norwegian Institute of International Affairs, 2014).

2. Ibid.

3. *Report of the Results of Radiological Study of Semipalatinsk Region During the Period 25 May–15 July 1957* (Moscow: Institute of Biophysics, 1957).

4. K. Gordeev et al., "Fallout from Nuclear Tests: Dosimetry in Kazakhstan," *Radiation and Environmental Biophysics* 41 (2002): 61–67, https://www.researchgate.net/publication/11358033_Fallout_from_nuclear_tests_Dosimetry_in_Kazakhstan.

5. Cynthia Werner and Kathleen Purvis-Roberts, *Unraveling the Secrets of the Past: Contested Versions of Nuclear Testing in the Soviet Republic of Kazakhstan* (Washington, DC: National Council for Eurasian and East European Research, 2005).

6. *Report of the Results of Radiological Study of Semipalatinsk Region.*

7. Bernd Grosche, "Semipalatinsk Test Site: Introduction," *Radiation and Environmental Physics* 41 (2002): 53–55.

8. Rob Edwards, "The Day the Sky Caught Fire," *New Scientist* (May 13, 1995), https://www.newscientist.com/article/mg14619772-300-the-day-the-sky-caught-fire/.

9. Gordeev et al., "Fallout from Nuclear Tests: Dosimetry in Kazakhstan."

10. Edwards, "The Day the Sky Caught Fire."

11. B. I. Gusev et al., "The Semipalatinsk Nuclear Test Site: A First Analysis of Solid Cancer Incidence (Selected Sites) Due to Test-Related Radiation," *Radiation and Environmental Biophysics* 37, no. 3 (1998): 209–14; Vakulchuk and Gjerde, *Semipalatinsk Nuclear Testing: The Humanitarian Consequences*; and V. F. Stepanenko, "Around Semipalatinsk Nuclear Test Site," *Journal of Radiation Research* (2006): A1–A13, https://www.ncbi.nlm.nih.gov/pubmed/16571923.

12. Fred Pearce, "After the Bomb," *New Scientist* (May 4, 2005), https://www.newscientist.com/article/mg18624986-400-interview-after-the-bomb/.

13. Hodge and Weinberger, *A Nuclear Family Vacation.*

14. Nick Paton Walsh, "When the Wind Blows," *Observer Magazine*, August 29, 1999.

15. Jacob Baynham, "From Russia with Radiation," *Slate*, September 2, 2013, http://www.slate.com/articles/news_and_politics/foreigners/2013/09/kazakhstan_was_site_of_the_soviet_union_s_first_atomic_bomb_the_kazak_people.html.

16. Vakulchuk and Gjerde, *Semipalatinsk Nuclear Testing.*

17. Daid Zardze et al., "Childhood Cancer Incidence in Relation to Distance from the Former Nuclear Testing Site in Semipalatinsk, Kazakhstan," *International Journal of Cancer* 59, no. 4 (November 15, 1994), doi: 10.1002/ijc.2910590407.

18. Aya Sakaguchi et al., "Radiological Situation in the Vicinity of Semipalatinsk Nuclear Test Site: Dolon, Mostik, Cheremushka and Budene Settlements," *Journal of Radiation Research* (2006): A101–A116, https://www.ncbi.nlm.nih.gov/pubmed/16571924.

19. Vakulchuk and Gjerde, *Semipalatinsk Nuclear Testing.*

20. Ibid.

21. Magdalena Stawkowski, "'I Am a Radioactive Mutant': Emergent Biological Subjectivities at Kazakhstan's Semipalatinsk Nuclear Test Site," *American Ethnologist* 43, no. 1 (February 2016), doi: 10.1111/amet.12269.

Chapter 6   Plutonium Mountain: Proliferation Paradise

1. Eben Harrell and David Hoffman, *Plutonium Mountain: Inside the 17-Year Mission to Secure a Dangerous Legacy of Soviet Nuclear Testing* (Cambridge, MA: Harvard University Press, 2013).

2. Ibid.

3. Siegfried Hecker, *Doomed to Cooperate: How American and Russian Scientists Joined Forces to Avert Some of the Greatest Post-Cold War Nuclear Dangers* (Los Alamos, NM: Bathtub Row Press, 2016), extracted at https://lab2lab.stanford.edu/sites/default/files/hecker_epilogue_vol_2.pdf.

4. Harrell and Hoffman, *Plutonium Mountain.*

5. Pearce, "After the Bomb."

6. Rachel Oswald, "High-Grade Plutonium Locked in Kazakhstan Mountain at Minimal Risk," Nuclear Threat Initiative, http://www.nti.org/gsn/article/high-grade-plutonium-locked-kazakhstan-mountain-minimal-risk/.

7. Harrell and Hoffman, *Plutonium Mountain*.

8. Ian Anderson, "Britain's Dirty Deeds at Maralinga," *New Scientist* (June 12, 1993), https://www.newscientist.com/article/mg13818772-700/.

9. Paul Langley, "Australia's Shame: Ignoring Its 'Black Mist' Atomic Radiation Victims," Anti-Nuclear.net, https://antinuclear.net/2010/04/21/australias-shame-ignoring-its-black-mist-atomic-radiation-victims/.

10. Alan Parkinson, "Maralinga: The Clean-Up of a Nuclear Test Site," *Medicine & Global Survival* 7, no. 2 (2002), http://www.ippnw.org/pdf/mgs/7-2-parkinson.pdf.

11. Anderson, "Britain's Dirty Deeds at Maralinga."

12. Patrick Cockburn, "Australia Keeps Wraps on UK Bomb Fallout Report," *Times* (London), July 5, 1989.

13. Parkinson, "Maralinga."

14. Candace Sutton, "Secret Outback Nuclear Testing Site Handed Back," *Mail Online*, November 6, 2014, http://www.dailymail.co.uk/news/article-2822906/.

Chapter 7   Mayak: "Pressed for Time" Behind the Urals

1. Kosenko, "Where Radiobiology Began in Russia."

2. "A Review of Criticality Accidents," Los Alamos National Laboratory, 2000, https://www.orau.org/ptp/Library/accidents/la-13638.pdf.

3. Mira Kosenko, *Analysis of Chronic Radiation Sickness Cases in the Population of the Southern Urals* (Bethesda, MD: Armed Forces Radiobiology Unit, 1994).

4. "A Review of Criticality Accidents," Los Alamos National Laboratory.

5. Mikhail Sokolnikov et al., "Lung, Liver and Bone Cancer Mortality in Mayak Workers," *International Journal of Cancer* 123 (2008), doi: 10.1002/ijc.23581.

6. Ethel Gilbert et al., "Liver Cancer in Mayak Workers," *Radiation Research* 154 (2000): 246–52.

7. Sergey Romanov et al., "Plutonium in the Respiratory Tract of Mayak Workers," Proceedings of the Ninth International Conference on Health Effects of Incorporated Radionuclides, IAEA, 2004, http://www.iaea.org/inis/collection/NCLCollectionStore/_Public/37/101/37101197.pdf.

8. Yuri Mokrov and G. Batorshin, "Experience in Eliminating the Consequences of the 1957 Accident at the Mayak Production Association," presented at International Experts' Meeting on Decommissioning and Remediation After a Nuclear Accident, IAEA, 2013, http://www-pub.iaea.org/iaeameetings/IEM4/Session2/Mokrov.pdf.

9. Thomas Cochran, *Russian/Soviet Nuclear Warhead Production* (Washington, DC: NRDC, 1992).

10. Kate Brown, *Plutopia: Nuclear Families, Atomic Cities, and the Great Soviet and American Plutonium Disasters* (Oxford, UK: Oxford University Press, 2013).

11. Kosenko, "Where Radiobiology Began in Russia."

12. Mokrov and Batorshin, "Experience in Eliminating the Consequences of the 1957 Accident at the Mayak Production Association."

13. Kosenko, "Where Radiobiology Began in Russia."

14. Brown, *Plutopia*.

15. Zhores Medvedev, "Two Decades of Dissidence," *New Scientist* (November 4, 1976), https://www.newscientist.com/article/dn10546-two-decades-of-dissidence/.

16. John Trabalka et al., "Analysis of the 1957–1958 Soviet Nuclear Accident," *Science* 209 (1980): 345–53, http://science.sciencemag.org/content/209/4454/345.

Chapter 8   Metlino: Even the Samovars Were Radioactive

1. Kosenko, *Analysis of Chronic Radiation Sickness Cases in the Population of the Southern Urals.*

2. Kosenko, "Where Radiobiology Began in Russia."

3. Ibid.

4. Kosenko, *Analysis of Chronic Radiation Sickness Cases in the Population of the Southern Urals.*

5. Kosenko, "Where Radiobiology Began in Russia."

6. Ibid.

7. Standring et al., "Overview of Dose Assessment Developments."

8. Sneve and Strand, *Regulatory Supervision of Legacy Sites.*

9. Alexander Akleyev, "Chronic Radiation Syndrome (CRS) in Residents of the Techa Riverside Villages," presentation to CONRAD conference, Munich, 2013, http://media.bsbb.de/Conrad/AKLEYEV.pdf.

10. Ljudmila Krestinina et al., "Leukaemia Incidence in the Techa River Cohort: 1953–2007," *British Journal of Cancer* 109, no. 11 (2013): 2886–93, https://www.ncbi.nlm.nih.gov/pmc/articles/PMC3844904/.

11. "Is Historic Soviet Radiation Data Too Hot to Handle?," *New Scientist* (December 7, 2016), https://www.newscientist.com/article/mg23231033–700.

12. Kosenko, *Analysis of Chronic Radiation Sickness Cases in the Population of the Southern Urals.*

13. "Mayak Plant's General Director Dismissed from His Post," Bellona, March 20, 2006, http://bellona.org/news/nuclear-issues/radwaste-storage-at-nuclear-fuel-cycle-plants-in-russia/2006–03-mayak-plants-general-director-dismissed-from-his-post.

14. Sneve and Strand, *Regulatory Supervision of Legacy Sites.*

15. Thomas Cochran et al., Radioactive Contamination at Chelyabinsk-65, Russia, *Annual Review of Energy and Environment* 18 (1993): 507–28, https://web.archive.org/web/20081209055500/http://docs.nrdc.org/nuclear/files/nuc_01009302a_112b.pdf.

16. Asker Aarkrog and Gennady Polikarpov, "Development of Radioecology in East and West," in *Radioecology and the Restoration of Radioactive-Contaminated Sites,* NATO ASI Series, ed. F. F. Luykx and M. J. Frissel (Dordrecht, Netherlands: Springer, 1996), 17–29, https://link.springer.com/chapter/10.1007/978–94–009–0301–2_2.

17. Charles Digges, "Environmentalists Skeptical About Russian Plans to Seal Off Radioactive Lake," Bellona, November 9, 2015, http://bellona.org/news/russian-human-rights-issues/russian-ngo-law/2015–11-environmentalists-skeptical-about-russian-plans-to-seal-off-radioactive-lake.

Chapter 9   Rocky Flats: Plutonium in the Snake Pit

1. "Facility History for Building 771 at the Rocky Flats Plant," M. H. Chew & Associates, April 1992, http://rockyflatsambushedgrandjury.com/wp-content/uploads/1992April-FacilityHistoryforBuilidng771attheRockyFlatsPlant.pdf.

2. Kristen Iversen, *Full Body Burden: Growing Up in the Nuclear Shadow of Rocky Flats* (London: Vintage, 2013).

3. "Facility History for Building 771 at the Rocky Flats Plant."

4. Thomas Cochran, *Overview of Rocky Flats Operations* (Washington, DC: NRDC, 1993).

5. Iversen, *Full Body Burden.*

6. "Facility History for Building 771 at the Rocky Flats Plant."

7. John E. Hill, interview by Dorothy D. Ciarlo, July 15, 1999, transcript, Maria Rogers Oral History Program, Boulder Public Library, Boulder, CO, http:// oralhistory.boulderlibrary.org/interview/oho998/.

8. "1957 Fire," Colorado State Government, https://www.colorado.gov/pacific /sites/default/files/HM_sf-rocky-flats-1957-fire.pdf.

9. Cochran, *Overview of Rocky Flats Operations.*

10. "1957 Fire," Colorado State Government.

11. "Atomic Plant Hit by $50,000 Fire," *St. Joseph Gazette* (CO), September 13, 1957, https://news.google.com/newspapers?id=mDhaAAAAIBAJ&sjid=QUwNAA AAIBAJ&pg=2955,4515771&dq=rocky-flats&hl=en.

12. Len Ackland, *Making a Real Killing: Rocky Flats and the Nuclear West* (Albuquerque: New Mexico University Press, 1999).

13. Iversen, *Full Body Burden.*

14. Wes McKinley and Caron Balkany, *The Ambushed Grand Jury: How the Justice Department Covered Up Government Nuclear Crimes and How We Caught Them Red-Handed* (New York: Apex, 2004).

15. Iversen, *Full Body Burden.*

16. Rocky Flats Project Office, *Rocky Flats Closure Legacy* (US Department of Energy, 2006), https://www.lm.doe.gov/Rocky_Flats_Closure.pdf.

17. Candelas Life, http://www.candelaslife.com/.

18. S. E. Poet and Ed Martell, "Plutonium-239 and Americium-241 Contamination in the Denver Area," *Health Physics* (October 1972), http:// journals.lww.com/health-physics/Abstract/1972/10000/Plutonium_239_and _Americium_241_Contamination_in.12.aspx.

19. John Aguilar, "Rocky Flats Stirs Strong Emotions, Pits Sides 10 Years After Cleanup," *Denver Post*, October 10, 2015, http://www.denverpost.com/2015/10/10 /rocky-flats-stirs-strong-emotions-pits-sides-10-years-after-cleanup/.

20. Poet and Martell, "Plutonium-239 and Americium-241 Contamination in the Denver Area."

21. "Plutonium in Breathable Form Found Near Rocky Flats," *Nuclear Monitor* 714 (2010), https://www.wiseinternational.org/nuclear-monitor/714/plutonium -breathable-form-found-near-rocky-flats.

22. Carl Johnson, "Cancer Incidence in an Area Contaminated with Radionuclides Near a Nuclear Installation," *Ambio* 10 (1981): 176–82, https://www .jstor.org/stable/4312671?seq=1#page_scan_tab_contents.

23. Carol Jensen, "Rocky Flats Downwinders Health Survey: Metropolitan State University of Denver," Rocky Flats Downwinders, 2016, http:// rockyflatsdownwinders.com/wp-content/uploads/2016/05/RFD-Health-Survey -Executive-Summary-Final.pdf.

Chapter 10   Colorado Silos: Uncle Sam's Nuclear Heartland

1. John LaForge, *Nuclear Heartland: A Guide to the 450 Land-Based Missiles of the United States* (Santa Fe, NM: Nukewatch, 2015).

2. "Missile Site Park," Weld County, CO, https://www.weldgov.com /departments/buildings_and_grounds/missile_site_park/.

3. James Brooke, "Counting the Missiles, Dreaming of Disarmament," *New York*

Times, March 19, 1997, http://www.nytimes.com/1997/03/19/us/counting-the
-missiles-dreaming-of-disarmament.html.

4. LaForge, *Nuclear Heartland*.

5. Andrew O'Hehir, "The Night We Almost Lost Arkansas," *Salon*, September
15, 2016, http://www.salon.com/2016/09/14/the-night-we-almost-lost-arkansas-a
-1980-nuclear-armageddon-that-almost-was/.

6. Jeannie Roberts, "Survivor Recalls 1965 Titan II Missile Silo Fire That Killed
53," *Arkansas Democrat-Gazette,* August 16, 2015, http://www.dailyrecord.com
/story/news/2015/08/16/survivor-recalls-titan-ii-missile-silo-fire-killed/31815507/.

7. Daniel Gross, "America's Missileers Stand Ready to Launch Nuclear
Weapons—and Pray They Won't Have To," Public Radio International, December
2, 2016, https://www.pri.org/stories/2016-12-02/americas-missileers-stand-ready
-launch-nuclear-weapons-and-pray-they-wont-have.

8. Dan Whipple, "Wyoming's Nuclear Might: Warren AFB in the Cold War,"
Wyoming State Historical Society, http://www.wyohistory.org/encyclopedia
/wyomings-nuclear-might-warren-afb-cold-war.

9. "Air Force Withheld Colorado Nuclear Missile Mishap from Pentagon Review
Team," Associated Press, January 22, 2016, http://gazette.com/air-force-withheld
-colorado-nuclear-missile-mishap-from-pentagon-review-team/article/1568375.

Chapter 11   Broken Arrows: *Dr. Strangelove* and the Radioactive Rabbits

1. David Philipps, "Decades Later, Sickness Among Airmen After a Hydrogen
Bomb Accident," *New York Times*, June 19, 2016, https://www.nytimes.com
/2016/06/20/us/decades-later-sickness-among-airmen-after-a-hydrogen-bomb
-accident.html?_r=0.

2. Gerry Hadden, "Palomares, the H-Bomb and Operation Moist Mop,"
Public Radio International, June 1, 2012, https://www.pri.org/stories/2012-06-01
/palomares-h-bomb-and-operation-moist-mop.

3. "US and Spain Agree to Clean Up Cold War Nuclear Accident," *Deutsche
Welle*, October 19, 2015, http://www.dw.com/en/us-and-spain-agree-to-clean-up
-cold-war-nuclear-accident/a-18792075.

4. European Commission, *Plutonium Contaminated Sites in the PALOMARES
Region*, 2010, https://ec.europa.eu/energy/sites/ener/files/documents/tech
_report_spain_palomares_2010_en.pdf.

5. Mats Eriksson, "On Weapons Plutonium in the Arctic Environment (Thule,
Greenland)," Riso National Laboratory, Roskilde, Denmark, 2002, http://www
.iaea.org/inis/collection/NCLCollectionStore/_Public/33/039/33039465.pdf.

6. Gordon Corera, "Mystery of Lost US Nuclear Bomb," *BBC News*, November
10, 2008, http://news.bbc.co.uk/1/hi/world/europe/7720049.stm.

Chapter 12   Windscale Fire: "A Cover-Up, Plain and Simple"

1. Brian Cathcart, *Test of Greatness: Britain's Struggle for the Atomic Bomb* (London:
John Murray, 1994).

2. Rossiter, *The Spy Who Changed the World*.

3. Ibid.

4. Paul Dwyer, "Windscale: A Nuclear Disaster," *BBC News*, October 5, 2007,
http://news.bbc.co.uk/1/hi/7030281.stm.

5. Fred Pearce, "Penney's Windscale Thoughts," *New Scientist* (January 7, 1988): 34–35; and UK Public Records Office, file reference AB 86/25.

6. UK Public Records Office, file reference AB 86/25.

7. "A Revised Transcript of the Proceedings of the Board of Inquiry into the Fire at Windscale Pile No. 1, October 1957," UK Atomic Energy Authority, 1989, accessed at http://news.bbc.co.uk/1/shared/bsp/hi/pdfs/05_10_07_ukaea.pdf; and Roy Herbert, "The Day the Reactor Caught Fire," *New Scientist* (October 14, 1982): 84–87.

8. David Lowry, "Obituary: Tom Tuohy," *Guardian*, May 7, 2008, https://www.theguardian.com/environment/2008/may/07/nuclearpower.

9. Harold Bolter, *Inside Sellafield: Taking the Lid off the World's Nuclear Dustbin* (London: Quartet Books, 1996).

10. Hunter Davies, ed., *Sellafield Stories: Life in Britain's First Nuclear Plant* (London: Constable & Robinson, 2012).

11. Ibid.

12. Bolter, *Inside Sellafield.*

13. Davies, *Sellafield Stories.*

14. "Windscale Fire," *New Scientist* (October 17, 1957): 7.

15. UK Public Records Office, file reference AB 86/25.

16. Ibid.; and Pearce, "Penney's Windscale Thoughts."

17. Lord Mills, "Windscale Atomic Plant Accident," *Hansard* 206, col. 467–75, November 21, 1957, http://hansard.millbanksystems.com/lords/1957/nov/21/windscale-atomic-plant-accident-1.

18. UK Public Records Office, file reference AB 86/25.

19. Dwyer, "Windscale: A Nuclear Disaster."

20. Fred Pearce, "Polonium Cloud Engulfs Windscale," *New Scientist* (March 31, 1983): 867; and John Urquhart, "Polonium: Windscale's Most Lethal Legacy," *New Scientist* (March 31, 1983): 873–74.

21. Fred Pearce, "Secrets of the Windscale Fire Revealed," *New Scientist* (September 29, 1983): 911.

22. "Estimates of Fission Product and Other Radioactive Releases from the 1957 Fire in Windscale Pile No. 1," Appendix IX, Lorna Arnold, *Windscale 1957: Anatomy of a Nuclear Accident* (London: Palgrave MacMillan, 1992).

23. M. J. Crick and G. S. Linsley, "An Assessment of the Radiological Impact of the Windscale Reactor Fire, October 1957," National Radiological Protection Board (UK) R-135, November 1982, https://www.osti.gov/opennet/servlets/purl/16291800/16291800.pdf.

Chapter 13   Three Mile Island: How Not to Run a Power Plant

1. Blowers, *The Legacy of Nuclear Power.*

2. "Three Mile Island Accident," *WNN*, http://www.world-nuclear.org/information-library/safety-and-security/safety-of-plants/three-mile-island-accident.aspx.

3. John Kemeny, President's Commission on the Accident at Three Mile Island, *Report of the President's Commission on the Accident at Three Mile Island* (Washington, DC: Library of Congress, 1979), http://www.threemileisland.org/downloads/188.pdf.

4. Peter Pringle and James Spigelman, *The Nuclear Barons* (London: Michael Joseph, 1982).

5. Bolter, *Inside Sellafield.*

6. Blowers, *The Legacy of Nuclear Power*.

7. Davies, *Sellafield Stories*.

8. Martin Gardner et al., "Results of Case-Control Study of Leukaemia and Lymphoma Among Young People Near Sellafield Nuclear Plant in West Cumbria," *British Medical Journal* 300 (1990): 423–29.

9. Heather Dickinson and Louise Parker, "Leukaemia and Non-Hodgkin's Lymphoma in Children of Male Sellafield Radiation Workers," *International Journal of Cancer* (2002), doi: 10.1002/ijc.10385.

10. Fred Pearce, "Massive Plutonium Levels Found in Cumbrian Corpses," *New Scientist* (August 14, 1986), https://www.newscientist.com/article/dn11666 -massive-plutonium-levels-found-in-cumbrian-corpses/.

11. James Randerson and Will Woodward, "Scientists Tested Plutonium Levels in Organs of Dead Sellafield Workers," *Guardian*, April 19, 2007, https://www .theguardian.com/uk/2007/apr/19/health.nuclear.

12. Davies, *Sellafield Stories*.

13. "Sellafield Radioactive Pigeon Scare," *BBC News*, February 11, 1998, http://news.bbc.co.uk/1/hi/uk/55612.stm.

14. Jay, *Britain's Atomic Factories*.

Chapter 14   Chernobyl: A "Beautiful" Disaster

1. Richard Mould, *Chernobyl Record: The Definitive History of the Chernobyl Catastrophe* (London: Institute of Physics, 2000).

2. Michael Bond, "Cheating Chernobyl," *New Scientist* (August 21, 2004), https://www.newscientist.com/article/mg18324615–300.

3. Ibid.

4. Svetlana Alexievich, *Voices from Chernobyl: Chronicle of the Future* (New York: Picador, 1997).

5. Mould, *Chernobyl Record*.

6. Milton Levenson, "A Different Chernobyl," *Doomed to Cooperate: US-Russian Lab-to-Lab Collaboration Story*, https://lab2lab.stanford.edu/e-archive/milton -levenson/different-chernobyl.

7. Mould, *Chernobyl Record*.

8. Ibid.

9. Ibid.

10. Ibid.

11. "Soviet Announces Accident at Electric Plant," *New York Times*, April 26, 1986, http://www.nytimes.com/learning/general/onthisday/big/0426.html; and Tom Wilkie, "The World's Worst Nuclear Accident," *New Scientist* (May 1, 1986): 17–19.

12. Don Higson, "Don't Compare Fukushima to Chernobyl," *New Scientist* (March 14, 2012), https://www.newscientist.com/article/mg21328566–500-dont -compare-fukushima-to-chernobyl/.

13. Allison, *Radiation and Reason*.

14. *Sources and Effects of Ionizing Radiation*, vol. 2 (New York: UNSCEAR, 2008), http://www.unscear.org/docs/publications/2008/UNSCEAR_2008_Annex-D -CORR.pdf.

15. Fred Pearce, "Chernobyl: The Political Fall-Out Continues," *UNESCO Courier*, October 2000.

16. Leyla Alyanak, *World Disasters Report 2000* (Geneva: IFRC, 2000), ch. 5.

17. V. K. Ivankov et al., "Leukemia Incidence in the Russian Cohort of Chernobyl Emergency Workers," *Radiation and Environmental Biophysics* (2012), doi: 10.1007/s00411–011–0400-y.

18. George Monbiot, "Evidence Meltdown," *Guardian*, April 5, 2011, http://www.monbiot.com/2011/04/04/evidence-meltdown/.

19. Pearce, "Chernobyl: The Political Fall-Out Continues."

20. *Sources and Effects of Ionizing Radiation*, vol. 1 (New York: UNSCEAR, 2000), http://www.unscear.org/docs/publications/2000/UNSCEAR_2000_Report_Vol.I.pdf.

21. Monbiot, "Evidence Meltdown."

22. Pearce, "Chernobyl: The Political Fall-Out Continues."

23. "The International Conference 'One Decade After Chernobyl: Summing up the Consequences of the Accident,'" IAEA, 1996, https://www.iaea.org/sites/default/files/infcirc510.pdf.

24. Alexievich, *Voices from Chernobyl*.

25. Mould, *Chernobyl Record*.

26. Alexievich, *Voices from Chernobyl*.

Chapter 15   Chernobyl: Vodka and Fallout

1. Thom Davies and Abel Polese, "Informality and Survival in Ukraine's Nuclear Landscape: Living with the Risks of Chernobyl," *Journal of Eurasian Studies* 6, no. 1 (2014): 34–45.

2. Norman Davies, *Europe: A History* (Oxford, UK: Oxford University Press, 1996).

3. Dmitri Bugai et al., "The Cooling Pond of the Chernobyl Nuclear Power Plant: A Groundwater Remediation Case History," *Water Resources Research* 33 (1997): 677–88, http://onlinelibrary.wiley.com/doi/10.1029/96WR03963/pdf.

4. Oleg Voitsekhovych et al., "Chernobyl Cooling Pond Remediation Strategy," IAEA, http://www-pub.iaea.org/iaeameetings/IEM4/29Jan/Voitsekhovych.pdf.

5. "Revive Chernobyl's Exclusion Zone: GCL-SI to Build PV Plant in Ukraine," GCL, 2017, http://en.gclsi.com/revive-chernobyls-exclusion-zone-gcl-si-to-build-pv-plant-in-ukraine/.

6. "Ukraine Preparing Contract to Return Spent Nuclear Fuel Recycling Products from Russia," *Interfax-Ukraine*, September 29, 2015, http://en.interfax.com.ua/news/general/293065.html.

Chapter 16   Chernobyl: Hunting in Packs

1. Tatiana Deryabina et al., "Long-Term Census Data Reveal Abundant Wildlife Populations at Chernobyl," *Current Biology* (October 5, 2015), http://www.cell.com/current-biology/abstract/S0960–9822(15)00988–4.

2. Vasyl Yoschenko et al., "Chronic Irradiation of Scots Pine Trees (*Pinus sylvestris*) in the Chernobyl Exclusion Zone: Dosimetry and Radiobiological Effects," *Health Physics* (October 2011), doi: 10.1097/HP.0b013e3182118094.

3. Anders Moller and Timothy Mousseau, "Biological Consequences of Chernobyl: 20 Years On," *Trends in Ecology and Evolution* (April 2006), doi: 10.1016/j.tree.2006.01.008.

4. Catherine Brahic, "Chernobyl-Based Birds Avoid Radioactive Nests," *New Scientist* (March 28, 2007), https://www.newscientist.com/article/dn11473 -chernobyl-based-birds-avoid-radioactive-nests/.

5. François Bréchignac et al., "Addressing Ecological Effects of Radiation on Populations and Ecosystems to Improve Protection of the Environment Against Radiation: Agreed Statements from a Consensus Symposium," *Journal of Environmental Radioactivity* (April 2016), doi: 10.1016/j.jenvrad.2016.03.021.

Chapter 17   Fukushima: A Scorpion's Discovery

1. "Fukushima Accident," *WNN*, http://www.world-nuclear.org/information -library/safety-and-security/safety-of-plants/fukushima-accident.aspx.

2. Wade Allison, "Man's Fear of Nuclear Technology Is Mistaken," paper based on lecture given at Second AGORA Conference, Tokyo, and British Chamber of Commerce in Japan, December 8 and 9, 2013, http://www.radiationandreason.com /uploads/enc_BetterThanFire.pdf.

3. "Fukushima: A Disaster 'Made in Japan,'" *WNN*, July 5, 2012, http://www .world-nuclear-news.org/RS_Fukushima_a_disaster_Made_in_Japan_050612I.html.

4. Allison, *Radiation and Reason.*

5. "Detectors Confirm Most Fuel Remains in Unit 2 Vessel," *WNN*, July 29, 2016, http://www.world-nuclear-news.org/RS-Detectors-confirm-most-fuel -remains-in-unit-2-vessel-2907164.html.

6. Anna Fifield and Yuki Oda, "Japanese Nuclear Plant Just Recorded an Astronomical Radiation Level. Should We Be Worried?," *Washington Post*, February 8, 2017, https://www.washingtonpost.com/news/worldviews/wp/2017/02/08 /japanese-nuclear-plant-just-recorded-an-astronomical-radiation-level-should-we -be-worried/?utm_term=.3a0dc686ba63.

7. "Ex-Worker During Fukushima Disaster Sues TEPCO, Kyushu Electric over Leukemia," *Kyodo News*, February 2, 2017, http://www.japantimes.co.jp /news/2017/02/02/national/crime-legal/ex-worker-fukushima-disaster-sues-tepco -kyushu-electric-leukemia/#.WVzn77pFyUl.

8. Justin McCurry, "Dying Robots and Failing Hope: Fukushima Clean-up Falters Six Years After Tsunami," *Guardian*, March 9, 2017, https://www .theguardian.com/world/2017/mar/09/fukushima-nuclear-cleanup-falters-six -years-after-tsunami.

Chapter 18   Fukushima: Baba's Homecoming

1. Donie O'Sullivan, "Photographer Sneaks into Fukushima Exclusion Zone," CNN.com, July 13, 2017, http://edition.cnn.com/2016/07/13/world/inside -fukushimas-radiation-zone/.

2. Isabel Reynolds, "Namie Radiation Evacuees Fear Return," *Japan Times*, October 26, 2016, http://www.japantimes.co.jp/news/2016/10/26/national/social -issues/namie-radiation-evacuees-fear-return/#.WVztiLpFyUl.

Chapter 19   Radiophobia: The Ghost at Fukushima

1. Geoff Brumfiel, "Fukushima Health-Survey Chief to Quit Post," *Nature* (February 20, 2014), http://www.nature.com/news/fukushima-health-survey-chief -to-quit-post-1.12463.

2. International Commission on Radiological Protection, "Protection of the Public in Situations of Prolonged Radiation Exposure," *Annals of the ICRP* 29, no. 1–2 (1999), http://www.icrp.org/publication.asp?id=ICRP%20Publication%2082.

3. Nick Jones, "Preparing for the Worst," *Red Cross Red Crescent*, 2016, http://www.rcrcmagazine.org/2016/04/preparing-for-the-worst/.

4. UN Scientific Committee on the Effects of Atomic Radiation, *Sources, Effects, and Risks of Ionizing Radiation* (New York: UNSCEAR, 2013), http://www.unscear.org/docs/reports/2013/13–85418_Report_2013_Annex_A.pdf.

5. Tetsuya Ohira et al., "Comparison of Childhood Thyroid Cancer Prevalence Among 3 Areas Based on External Radiation Dose After the Fukushima Daiichi Nuclear Power Plant Accident: The Fukushima Health Management Survey," *Medicine* (August 2016), doi: 10.1097/MD.0000000000004472.

6. Hyeong Ahn, "South Korea's Thyroid-Cancer 'Epidemic'—Turning the Tide," *New England Journal of Medicine* 373 (2015): 2389–90.

7. Toshihide Tsuda et al., "Thyroid Cancer Detection by Ultrasound among Residents Ages 18 Years and Younger in Fukushima, Japan: 2011 to 2014," *Epidemiology* 27 (2016): 316–22.

8. Masaharu Maeda et al., "Fukushima, Mental Health and Suicide," editorial, *Journal of Epidemiology and Community Health* (2015), doi: 10.1136/jech-2015–207086.

9. Yasuto Kunii et al., "Severe Psychological Distress of Evacuees in Evacuation Zone Caused by the Fukushima Daiichi Nuclear Power Plant Accident," *PLOS* (July 2016), doi: 10.1371/journal.pone.0158821.

10. Ryoko Ando, "Reclaiming Our Lives in the Wake of a Nuclear Plant Accident," *Clinical Oncology* 28 (January 2016): 275–76.

11. Geoff Brumfiel, "Fukushima: Fallout of Fear," *Nature* (January 16, 2013), http://www.nature.com/news/fukushima-fallout-of-fear-1.12194.

12. Shuhei Nomura et al., "Post-Nuclear Disaster Evacuation and Survival Amongst Elderly People in Fukushima," *Preventive Medicine* 82 (2016): 77–82.

13. Claire Leppold et al., "Public Health After a Nuclear Disaster: Beyond Radiation Risks," *Bulletin of WHO* 94 (2016), https://www.ncbi.nlm.nih.gov/pmc/articles/PMC5096345/.

Chapter 20   Millisieverts: A Dose of Reason

1. Adrienne Crezo, "9 Ways People Used Radium Before We Understood the Risks," *Mental Floss*, http://mentalfloss.com/article/12732/9-ways-people-used-radium-we-understood-risks.

2. Prentiss Orr, "Eben M. Byers: The Effect of Gamma Rays on Amateur Golf, Modern Medicine and the FDA," *Allegheny Cemetery Heritage* (Fall 2004), http://www.alleghenycemetery.com/images/newsletter/newsletter_XIII_1.pdf.

3. "Milk River," *Visit Jamaica*, http://www.visitjamaica.com/milk-river.

4. IAEA, "The Radiological Accident in Goiania," 1988, http://www-pub.iaea.org/MTCD/publications/PDF/Pub815_web.pdf.

5. Krishnan Nair et al., "Population Study in the High Natural Background Radiation Area in Kerala, India," *Radiation Research* 152 (1999): 145–48.

6. Kent Tobiska et al., "Global Real-Time Dose Measurements Using the Automated Radiation Measurements for Aerospace Safety (ARMAS) System," *Space Weather* (2016), doi: 10.1002/2016SW001419.

7. ICRP, "Protection of the Public in Situations of Prolonged Radiation Exposure."

8. Karen Smith, "Overview of Radiological Dose and Risk Assessment," IAEA, 2011, https://www.iaea.org/OurWork/ST/NE/NEFW/documents/IDN/ANL%20 Course/Day_5/RiskOverview_revised.pdf.

9. "Fact Sheet on Fallout Report and Related Maps," Institute of Energy and Environmental Research, https://ieer.org/resource/factsheets/fact-sheet-fallout -report-related/.

10. *Report to UN General Assembly 1958* (New York: UNSCEAR), http://www .unscear.org/docs/publications/1958/UNSCEAR_1958_Report.pdf.

11. ICRP, "Protection of the Public in Situations of Prolonged Radiation Exposure."

12. Yehoshua Socol and Ludwik Dobrzynski, "Atomic Bomb Survivors Life-Span Study," *Dose-Response* (2015), 10.2203/dose-response.14–034.Socol.

13. Bill Sacks et al., "Epidemiology Without Biology: False Paradigms, Unfounded Assumptions, and Specious Statistics in Radiation Science," *Biological Theory* 11 (2016): 69–101.

14. Wade Allison, "Nuclear Energy and Society, Radiation and Life— The Evidence," presented at Oxford Energy Colloquium, Oxford University, November 1, 2016, https://www.researchgate.net/publication/311175620_Nuclear _energy_and_society_radiation_and_life_-_the_evidence_1.

15. Arthur Tamplin and Thomas Cochran, "Radiation Standards for Hot Particles," NRDC, 1974, http://www.vff-marenostrum.org/Nuntium-Novitatum /PDF/Tamplin.Cochran.Radiation.std.for.hot.particles.pdf.

16. Brian Flowers, *Nuclear Power and the Environment* (London: Royal Commission on Environmental Pollution, 1976).

17. William Moss and Roger Eckhardt, "The Human Plutonium Injection Experiments," *Los Alamos Science* 23 (1995), https://fas.org/sgp/othergov/doe/lanl /pubs/00326640.pdf.

18. Bernard Cohen, "The Myth of Plutonium Toxicity," *Nuclear Energy*, 1985, https://web.archive.org/web/20110806235718/http:/russp.org/BLC-3.html.

Chapter 21   Sizewell: The Nuclear Laundryman

1. Dan Gould, "Night Falls on Sizewell A," *Nuclear Engineering International*, April 2, 2007, http://www.neimagazine.com/features/featurenight-falls-on-sizewell-a/.

2. "Old Nuclear Event in the Open," *WNN*, June 12, 2009, http://www.world -nuclear-news.org/rs_old_nuclear_event_in_the_open_1206091.html.

3. *Uncontrolled Partial Emptying of the Sizewell A Irradiated Fuel Cooling Pond, 7th January 2007*, Project Assessment Report 2007/0011, Nuclear Installations Inspectorate.

4. "Old Nuclear Event in the Open," *WNN*, June 12, 2009, http://www.world -nuclear-news.org/RS_Old_nuclear_event_in_the_open_1206091.html.

5. Large and Associates, "Sizewell A—Cooling Pond Recirculation Pipe Failure Incident of 7 January 2007 Assessment of the NII Decision Making Process," http:// www.largeassociates.com/LA%20reports%20&%20papers/3179%20Sizewell%20A%20 Pond%20Drainage/R3179-A3.pdf.

6. Louise Gray, "Nuclear Disaster Averted by Dirty Laundry," *Daily*

Telegraph, June 11, 2008, http://www.telegraph.co.uk/news/earth/energy
/nuclearpower/5509277/Nuclear-disaster-averted-by-dirty-laundry.html.

7. Rob Pavey, "Reactors Sealed for Last Time," *Augusta Chronicle*, June 29, 2011,
http://chronicle.augusta.com/metro/2011-06-29/reactors-sealed-last-time.

8. Matthew Wald, "Dismantling Nuclear Reactors," *Scientific American* (January
26, 2009), https://www.scientificamerican.com/article/dismantling-nuclear/.

9. "U.S. Utility's Deferred Reactor Clean-Up Shows Cost Pressure on Early
Closures," *Nuclear Energy Insider*, September 21, 2016, http://analysis.nuclearenergy
insider.com/us-utilitys-deferred-reactor-clean-shows-cost-pressure-early-closures.

10. Christoph Steitz and Barbara Lewis , "EU Short of 118 Billion Euros in
Nuclear Decommissioning Funds—Draft," Reuters, February 16, 2016, http://uk
.reuters.com/article/uk-europe-nuclear-idUKKCN0VP2KN.

11. "France's Lower-Cost Decommissioning Plan Rests on Chooz A
Reactor Learnings," *Nuclear Energy Insider*, February 22, 2017, http://analysis.
nuclearenergyinsider.com/frances-lower-cost-decommissioning-plan-rests-chooz
-reactor-learnings.

12. "Closing and Decommissioning Nuclear Reactors," *UNEP Year Book 2012*,
http://staging.unep.org/yearbook/2012/pdfs/UYB_2012_CH_3.pdf.

13. "Uranium Incident—No-one Hurt," *Burnham Advertiser*, February 1967,
accessed at http://turpidity.blogspot.co.uk/2006/06/nuclear-incident-no-one
-hurt.html.

14. "Graphite Research to Support AGR Life Extensions," *WNN*, February 22,
2016, http://www.world-nuclear-news.org/C-Graphite-research-to-support-AGR
-life-extensions-2202164.html.

15. G. Holt, "Radioactive Graphite Management at UK Magnox Nuclear
Power Stations," IAEA, 2001, http://www-pub.iaea.org/MTCD/publications/PDF
/ngwm-cd/PDF-Files/paper%2017%20(Holt).pdf.

Chapter 22   Sellafield: Stone Circles and Nuclear Legacies

1. "Nuclear Provision: The Cost of Cleaning Up Britain's Historic Nuclear
Sites," Nuclear Decommissioning Authority, https://www.pgov.uk/government
/publications/nuclear-provision-explaining-the-cost-of-cleaning-up-britains-nuclear
-legacy/nuclear-provision-explaining-the-cost-of-cleaning-up-britains-nuclear
-legacy, accessed July 6, 2017,

2. Bolter, *Inside Sellafield*.

3. "Why the Future Will Be Different," Chris Huhne, lecture to Royal Society,
October 13, 2011, https://www.gov.uk/government/speeches/the-rt-hon-chris
-huhne-mp-speech-to-the-royal-society-why-the-future-of-nuclear-power-will
-be-different.

4. Bolter, *Inside Sellafield*.

5. "Decommissioning Milestone Achieved at Pile Fuel Storage Pond," *WNN*,
March 2, 2016, http://www.world-nuclear-news.org/WR-Decommissioning
-milestone-achieved-at-Pile-Fuel-Storage-Pond-02031602.html.

6. "Radioactive Sludge Removed from UK's Pile Fuel Storage Pond," *WNN*,
December 22, 2016, http://www.world-nuclear-news.org/WR-Radioactive-sludge
-removed-from-UKs-Pile-Fuel-Storage-Pond-22121601.html; and "First Sellafield
Pond Fuel Sludge Encapsulated," *WNN*, February 20, 2017, http://www.world
-nuclear-news.org/WR-First-Sellafield-fuel-pond-sludge-encapsulated-2002174.html.

7. "Doors Installed at Sellafield Silo," *WNN*, December 9, 2016, http://www
.world-nuclear-news.org/WR-Doors-installed-at-Sellafield-silo-09121601.html.

8. Bolter, *Inside Sellafield*.

9. Ibid.; and Walt Patterson, *Going Critical* (London: Paladin, 1985).

10. "Sellafield Nuclear Waste Storage Is 'Intolerable Risk,'" *BBC News*,
November 7, 2012, http://www.bbc.co.uk/news/uk-england-cumbria-20228176.

11. Flowers, *Nuclear Power and the Environment*.

12. Gordon Thompson, "Radiological Risk at Nuclear Fuel Reprocessing Plants,"
version 2, Institute for Resource and Security Studies (July 2014), http://www.academia
.edu/12471352/Radiological_Risk_at_Nuclear_Fuel_Reprocessing_Plants_2014.

13. "The Storage of Liquid High Level Waste at Sellafield: Revised Regulatory
Strategy," Office of Nuclear Regulation, 2011, http://www.onr.org.uk/halstock
-sellafield-public.pdf.

Chapter 23   Hanford: Decommissioning an Industry

1. David Gutman, "Thousands of Hanford Workers Take Cover After Cave-in
of Tunnel with Radioactive Waste," *Seattle Times*, May 9, 2017, http://www
.seattletimes.com/seattle-news/environment/hanford-declares-alert-emergency
-evacuates-workers-because-of-problems-with-contaminated-tunnels/.

2. Michele Stenehjem Gerber, *On the Home Front: The Cold War Legacy of the
Hanford Nuclear Site* (Lincoln, NE: Bison Books, 1992).

3. "Hanford's Tank Waste," Hanford Challenge, http://www.hanfordchallenge
.org/tank-waste/.

4. Annette Cary, "GAO Pushes Cheaper Way to Treat Hanford Radioactive
Tank Waste," *Tri-City Herald*, May 2, 2017, http://www.tri-cityherald.com/news
/local/hanford/article148521069.html.

5. Ralph Vartabedian, "Hanford Nuclear Weapons Site Whistle-Blower Wins
$4.1 Million Settlement," *Los Angeles Times*, August 13, 2015, http://www.latimes
.com/nation/la-na-hanford-whistleblower-settlement-20150813-story.html.

6. "Hanford Site Clean-up Completion Framework," US Department of
Energy, 2013, http://www.hanford.gov/page.cfm/HanfordSiteCleanupCompletion
Framework.

7. Blowers, *The Legacy of Nuclear Power*.

8. "Tenders for Russian Submarine Fuel Removal," *WNN*, February 7, 2014,
http://www.world-nuclear-news.org/WR-Tenders-for-Russian-submarine-fuel
-removal-0702144.html; and "First Used Fuel Shipment Leave Andreeva Bay,"
WNN, June 27, 2017, http://www.world-nuclear-news.org/WR-First-used-fuel
-shipment-leaves-Andreeva-Bay-2806177.html.

9. Thomas Nilsen, "Radiation Researchers Denied Access to Naval Waters,"
Barents Observer, April 30, 2014, http://barentsobserver.com/en/security/2014/04
/radiation-researchers-denied-access-naval-waters-30-04.

10. Yaroslava Kiryukhina, "Eyewitness: Tragedy of Soviet Nuclear Submarine
K-27," *BBC News*, January 24, 2013, http://www.bbc.co.uk/news/world
-europe-21148434.

11. Charles Digges, "Russia Announces Enormous Finds of Radioactive Waste
and Nuclear Reactors in Arctic Seas," Bellona, August 28, 2012, http://bellona.org
/news/nuclear-issues/radioactive-waste-and-spent-nuclear-fuel/2012–08-russia
-announces-enormous-finds-of-radioactive-waste-and-nuclear-reactors-in-arctic-seas.

12. *Matters Related to the Disposal of Radioactive Waste at Sea*, report to International Maritime Organization, September 14, 1993, http://www.atomicreporters.com/wp-content/uploads/2013/10/lc-16_inf-2-part-1.pdf; and "Russian Submersibles to Monitor K-278 Nuclear Sub Disaster Area," *Sputnik*, August 3, 2007, https://sputniknews.com/russia/200708037031 5960/.

13. Thomas Nilsen, "These Dangerous Radioactive Reactors to Be Lifted Off the Waters," *Barents Observer*, January 30, 2017, https://thebarentsobserver.com/en/security/2017/01/these-dangerous-radioactive-reactors-soon-be-lifted-waters.

14. Jonathan Morris, "Devonport: Living Next to a Nuclear Submarine Graveyard," *BBC News*, October 2, 2014, http://www.bbc.co.uk/news/uk-england-devon-28157707.

15. Jonathan Morris, "Laid-Up Nuclear Submarines at Rosyth and Devonport Cost £16m," *BBC News*, June 3, 2015, http://www.bbc.co.uk/news/uk-england-devon-32086030.

16. "MoD Planned to Dump Old Nuclear Submarines in Sea," *Herald* (Scotland), August 19, 2012, http://www.heraldscotland.com/news/13069799.MoD_planned_to_dump_old_nuclear_submarines_in_sea/.

Chapter 24   Gorleben: Passport to a Non-Nuclear Future?

1. Michael Frohlingsdorf et al., "Germany's Homemade Nuclear Waste Disaster," *Der Spiegel*, February 21, 2013, http://www.spiegel.de/international/germany/germany-weighs-options-for-handling-nuclear-waste-in-asse-mine-a-884523.html.

2. Ingrid Lowin and Werner Lowin, *Gorleben XXL* (Luchow: Druck- und Verlagsgesellschaft, 2010).

3. *Report of the German Commission on the Storage of High-Level Radioactive Waste*, translation, 2016, http://www.nuclear-transparency-watch.eu/wp-content/uploads/2017/02/Summary_Report-of-the-German-Commission-on-the-Storage-of-High-Level-Radioactive-Waste_EN.pdf.

4. Blowers, *The Legacy of Nuclear Power*.

Chapter 25   Waste: Out of Harm's Way

1. "Inventory of Radioactive Waste Disposals at Sea," IAEA, 1999, http://www-pub.iaea.org/MTCD/Publications/PDF/te_1105_prn.pdf.

2. Ralph Vartabedian, "Nuclear Accident in New Mexico Ranks Among the Costliest in U.S. History," *Los Angeles Times*, August 22, 2016, http://www.latimes.com/nation/la-na-new-mexico-nuclear-dump-20160819-snap-story.html.

3. "Storage and Disposal of Spent Fuel and High Level Radioactive Waste," IAEA, https://www.iaea.org/About/Policy/GC/GC50/GC50InfDocuments/English/gc50inf-3-att5_en.pdf, accessed October 4, 2017.

4. Bruce Finley, "Colorado and Nation Face 70,000-Ton Nuclear Waste Burden," *Denver Post*, May 24, 2016, http://www.denverpost.com/2016/05/24/feds-favor-mini-nuke-power-plants-but-still-face-70k-ton-disposal-burden/; and Ahmed Sharif, "Spent Nuclear Fuel in the U.S., France and Finland," Stanford University, 2011, http://large.stanford.edu/courses/2011/ph241/sharif1/.

5. Paul Brown, "Kazakhstan Reveals Solution to Its Nuclear Waste Crisis: Import More," *Guardian*, November 21, 2002, https://www.theguardian.com/environment/2002/nov/21/internationalnews.energy.

6. "Going Nuclear in South Australia?," *WNN*, November 18, 2016, http://www.world-nuclear-news.org/V-Going-nuclear-in-South-Australia-18111601.html.

7. Edwin Cartlidge, "Plutonium Plans in Limbo," *Nature* (August 8, 2011), http://www.nature.com/news/2011/110808/full/476140a.html.

8. "Obama Seeks to Terminate Mox Project at Savannah River," *WNN*, February 10, 2016, http://www.world-nuclear-news.org/UF-Obama-seeks-to-terminate-MOX-project-at-Savannah-River-10021601.html.

9. "Russia Suspends Plutonium Agreement with USA," *WNN*, October 4, 2016, http://www.world-nuclear-news.org/NP-Russia-suspends-plutonium-agreement-with-USA-04101601.html.

10. "Managing the UK Plutonium Stockpile," Parliamentary Office of Science and Technology, 2016, http://researchbriefings.files.parliament.uk/documents/POST-PN-0531/POST-PN-0531.pdf.

11. Duncan Clark, "New Generation of Nuclear Reactors Could Consume Radioactive Waste as Fuel," *The Guardian*, February 2, 2012, https://www.theguardian.com/environment/2012/feb/02/nuclear-reactors-consume-radioactive-waste.

12. Geoffrey Boulton, *Strategy Options for the UK's Separated Plutonium* (London: Royal Society, 2007), https://royalsociety.org/~/media/Royal_Society_Content/policy/publications/2007/8018.pdf.

13. Bolter, *Inside Sellafield*.

14. Boulton, *Strategy Options for the UK's Separated Plutonium*.

Conclusion   Making Peace in Nagasaki

1. Mark Lynas, *The God Species: Saving the Planet in the Age of Humans* (London: Fourth Estate, 2011).

2. "Korea's Nuclear Phase-out Policy Takes Shape," *WNN*, June 19, 2017, http://www.world-nuclear-news.org/NP-Koreas-nuclear-phase-out-policy-takes-shape-1906174.html.

3. Jungk, *Brighter Than a Thousand Suns*.

4. Barack Obama, remarks in Prague, April 5, 2009, available at http://www.ploughshares.org/sites/default/files/newss/Palm%20Sunday%20Speech.pdf?_ga=1.193905327.36620478.1489416123.

5. Takashi Nagai, *Atomic Bomb Rescue and Relief Report* (Nagasaki: Nagasaki Association for Hibakushas' Medical Care, 2000); and Raisuke Shirabe, *A Physician's Diary of the Atomic Bombing and Its Aftermath* (Nagasaki: Nagasaki Association for Hibakushas' Medical Care, 2002).

6. "Nagasaki Mayor Urges World to Use Collective Wisdom to Abolish Nuclear Arms," *Kyodo News*, August 9, 2016, http://www.japantimes.co.jp/news/2016/08/09/national/nagasaki-mayor-urges-world-use-collective-wisdom-abolish-nuclear-weapons/#.WV4dtLpFyUl.

7. Joe Cirincione, "The Historic UN Vote for Banning Nuclear Weapons," *Huffington Post*, October 31, 2016, http://www.huffingtonpost.com/joe-cirincione/historic-un-vote-on-banni_b_12679132.html.

8. "Text of President Obama's Speech in Hiroshima, Japan," *New York Times*, May 27, 2016, https://www.nytimes.com/2016/05/28/world/asia/text-of-president-obamas-speech-in-hiroshima-japan.html.

Index